CRG Bologna - Ph.F.Fontana
Castelli SpA
Azienda certificata ISO 9001 o UNI-EN 29001
Via Olmatello, 21 - 40064 Ozzano dell'Emilia (BO)
ITALIA - Tel. 051/6513711 - Fax 051/6513799

3D
L'UNICO
SISTEMA
PER
UFFICI...
VERAMENTE
APERTI
CASTELLI®
Arredamenti per l'ufficio e la collettività

Piroscafo
Luca Meda
Molteni & C. 20034 Giussano Telefono: Telex: Telefax:
Via Rossini 50 (Milano) Italia (0362) 851334 330188 (0362) 354448

Molteni & C
M

PRS Public Relations System (BO) - Photo/Art Barut

Foto: Matteo Tresoldi

galleria Borghese

WITH COMPLIMENTS

La pubblicazione di questa rivista
è possibile
grazie al sostegno economico
e organizzativo delle aziende:

The publication of this magazine
is possible
with the sponsorship of the
following firms:

La publication de cette revue
est possible grâce au soutien
économique et à la capacité
d'organisation des entreprises:

Ermöglicht wird die
Veröffentlichung
dieser Zeitschrift
dank der finanziellen und
organisatorischen Unterstützung
der Firmen:

Abet Laminati
B&B Italia
Castelli
Molteni & C.
Sabiem

Con la collaborazione di
With the collaboration of
Avec la collaboration de
Unter Mitwirkung von

ENEL

A. Sant'Elia, *La centrale elettrica*, 1914.
From: *Antonio Sant'Elia. L'architettura disegnata* (Venice: Marsilio, 1991).

(Electricity. United States and Ussr, France and Italy)

Rassegna
Themes in Architecture

Editor
Vittorio Gregotti

Editorial Staff
Pierluigi Cerri
Dario Matteoni
Livia Piperno

Editorial Coordinator
Maresin Cavagna

Art Director
Pierluigi Cerri

Production
Maurizio Zanuso

Translations
Language Consulting
Antonella Bergamin
Cioni Carpi
Amedeo Mercurio

Collaborators
Isabella Fantin
Francesca Molteni

This issue edited by Jacques Gubler and Sara Protason

Quarterly
year XVII, 63 - 1995/III
Publisher: CIPIA srl
via Stalingrado 97/2, 40128 Bologna, Italy
Editorial Offices: via Matteo Bandello 20
20123 Milano, Italy, tel. (2) 481.41.41-481.24.48
fax (2) 481.41.43-481.24.48
Publishing Services: Editrice Compositori srl
via Stalingrado 97/2, 40128 Bologna, Italy
tel. (51) 32.78.11, fax (51) 32.78.77

Registered in Milan, Italy
Price: Lire 45,000
Back issues: Lire 49,000
Subscription: one year (four issues)
Lire 156,000
Advertising representative:
Editrice Compositori srl,
via Stalingrado 97/2, 40128 Bologna, Italy
tel. (51) 32.78.11, fax (51) 32.78.77
Distribution (Italy, foreign and subscriptions):
CIPIA, via Stalingrado 97/2,
40128 Bologna, Italy
tel. (51) 32.79.29, fax (51) 32.78.77
Color by Graphicolor srl, Milano
Printed by Tipografia Compositori srl, Bologna

We would like to thank for their precious collaboration Piero Castellini Baldissera, Luisa Castellini Portaluppi, Maria Antonietta Di Marco (ENEL), Lorenzo Pucci (ENEL), Onofrio Seregni (ENEL, Milan), Alessandra Zoppi, A. Chernikov Archives in Moscow, the Azienda Energetica Municipale in Turin. Special thanks to Luciana Ravanel and Maurice Culot for their valuable suggestions.

Jacques Gubler

Introduction

Materiality and Immateriality of Electricity

Etymologically speaking electricity is allegoric. The allegory does not lend its body to an abstract, composite, complex idea, such as Republic, Justice, Fortune. In modern physics, electricity is translated into pretty equations which refer to the energy contained in the corpuscular structure of the material.

At the same time electricity occupies various "allegoric architectural bodies" which witness its physical presence in the geography of the city and the land. On the one side the abstract concept of the network is conveyed by the diagram codification, on the other the network is materialized as a series of technical works such as, for example, electric household appliances which we recognize since childhood.

We know that children are great consumers of "light", of "torches", of "spotlights." In turn, the constitution of product associated with the diffusion of these objects entails a significant abstract element. The rates and codes of the proverbial "electricity bill" are the signs of a commercial management whose parameters, since a century, shed luster to the tertiary service sector, in the same way as banks and tourism. In fact, the concept of industrial services, and its water-gas-electricity trilogy, was developed in the second half of the 19th century to support both primary sectors (agriculture) and secondary ones (the industry transforming raw materials).

Electricity was therefore to play a central role in the galaxy of the capitalistic world. Its organizational abilities and the dependency it subjected other industrial constellations to, made it the first, and at the same time the last, in terms of importance.

The metaphor of the nervous system is a fitting one: indeed, it operates in a constant interaction between the center and the periphery. *Ergo* its characteristic dual situation of omnipresence and fragility, of predominance and abandonment to sudden stupor.

In his poem *La Chambre Double* Charles Baudelaire depicts this unstable reality: a room full of every seduction aimed at desire and pleasure suddenly becomes a place of repulsion and misery.

This poetic theme can also be found in the *Grande Panne de Lumière* which evokes both the scenario of the catastrophe suddenly disrupting everyday order and the rational technique of the *coup d'état* when the armed group seizes, in the first place, the places from where the infrastructure is controlled.

The authors of the monumental *Storia dell'elettricità in Italia* have studied the economic and cultural associations of the phenomenon. Published before the structural crises of the 1990s, the statistic of production is particularly eloquent: it aims at demonstrating that throughout a century one of the constants of electricity has been that of an annual growth in consumption, a growth which has not been interrupted either by the two world wars or the "historical" economic crises. The recent innovations in the field of miniaturization and the superconductivity of mechanical components will not be examined in this issue of *Rassegna*, whose *ante quem* term is found in the late 1930s, and in other words before the development of radar and computers.

1

Reason Versus Magic, or When Theory Precedes Practice

Without wanting to play with words, we note that electricity occupied an emblematic place in the drawing room of the second half of the 18th century, during the Age of Enlightenment. An interest in scientific experiments, whether it is a matter of trying the sharp shock of the discharge on one's own flesh or observing the muscular contractions of a dead frog connected to electrodes, is an invitation to explanations.

Physicists are philosophers. They write, debate, are asked to comment their discoveries. Did not Mary Shelley Wollstonecraft owe her inspiration for the subject of the creation of a living machine to this theoretic power? From 1818 to the end of the 20th century one was often to remember that the experiment carried out by Dr. Frankenstein, whose visit card carried the title of "Modern Prometheus," was only partially successful.

It is to romanticism that we owe the first poetic exploration of electricity. In the 1850s for Whitman the "body electric" instigated the gymnastics of words which places under tension all sensations capable of testing the inventory of sentiments. To Baudelaire, in the same years, modernity was attained by the immersion of the individual in the stream of the big city's crowds: the poet also becomes an actor and a spectator. He acts and becomes dispersive. At the same time he observes and flies over the crowd. The adjective "electric", then, suggests the quickness and sensual stimuli caused by the large number of encounters. We do not have to wait for the poetry of the canonical avant-gardes (Apollinaire, Cendrars, Marinetti, Cocteau) in order for electricity to play its role as a detonator of imagery.

Throughout the 19th century the electric flow aided bazaar magicians who made the occult powers of "magnetism" popular. We are poles apart from the Enlightenment. A mysterious, para- (from Greek) or super-natural world was supposed to exist.

This tendency to encourage credulity, an often profitable credulity, was denounced in 1858 by Carlo Collodi in a memorable text titled *Puhm! Puhm! Puhm!, comincia l'esame.* The author of Pinocchio imagines a dialogue which has to be memorized as the preparation for the admission exam of free will. The text summarizes the antinomy of magic and reason: since this antinomy is focused on electricity and since Collodi's irony was conceived a century after the *Letters on Electricity* by Benjamin Franklin, a key personality in the popular diffusion of physics, it is worthwhile to quote a long passage:

What is magnetism?
It is the art of making others daydream...
Do you believe in magnetism?
As much as canon Petrarca believed in the virginity of Donna Laura.
What is the magnetic fluid?
It's an expression invented especially in order not to be understood.
What is the great miracle of the magnetic fluid?
That of making the charlatans in the street appear as so many Professors in physics.
How does the magnetic fluid work?
It closes the mouth of suckers who do not know what to answer.
What is a magnetized person?
An individual of either sex, who begins offering

1. The "spark" at the Palais de l'Electricité, Exposition Internationale des Arts et des Techniques, Paris, 1937. Photothèque EDF, Paris. Photo Brassai.

2. Transformers in an outdoor plant. Banca Immagini ENEL.

him or herself to be magnetized as a joke,
and who ends up believing that he or she has really been magnetized.
What is the positive aspect of the magnetizing experiments?
The entrance ticket. [...]
What does magnetism really consist of?
Of the good will of the spectators.
What is required in order to become a good magnetizer?
Great impudence to make others believe what we ourselves do not believe.[1]

Collodi's battle against the modern quacks rejoined Benjamin Franklin's previous efforts of clarification at the service of public instruction. Franklin has been handed down to posterity as the inventor of the lightning-conductor: but we tend to forget that his discovery was violently contested by the advocates of a method consisting of ringing the church bells in order to drive the lightning away from the homes. A recent book by Bernard Cohen reminds us of the importance of the scientific work of the physicist of Philadelphia, founder of the University of Pennsylvania.[2] The author is astonished that philological research has forgotten this aspect of Franklin's work and has chosen to focus on the mass of writings on politics, morality, economics and practical issues, describing domestic life. In his capacity of *Paterfamilias* of the United States, Franklin represents, in the school books, the value of "good sense." He is identified with pragmatism and with enterprise; "good sense" must produce a "good mood" through work, which leads to moral and material enhancement. This approach, retrospective and patriotic, ignores that the concept of "good sense" was, in the 18th century, associated with the system of Cartesian rationalism.

Cohen remembers that the *Experiences & Observations on Electricity Made at Philadelphia (1751-1754)*, which were immediately translated into French, were to diffuse the theory of positive and negative electricity, associated with the hypothesis of the "conservation of the charge," in Europe.

Inspired by Newton's research, Franklin confirms the thought of Bacon: that Pure Science and the theory of Nature precede a mechanical and empiric application of the principles. Thus the experiment of the kite which captures the lightning, which is definitely difficult to realize, had been described in literature before it was put into practice. A century later English physicist George Chrystal recalls the final moment of the experiment: "In company of his son, Franklin raised the kite like a

2

common one, in the first thunderstorm which happened in the month of June 1752. He found that the clouds were sometimes positively, sometimes negatively electrified, and so demonstrated the perfect identity of lightning and electricity."[3] The application which was developed on the basis of this, for the capture and conduction of lightning, was undoubtedly to be the first application of the theory of electricity in architecture. And the invention was an ingenious and useful one, in a century when fires often burned quite vast areas. Today it may seem ridiculous to us that Franklin's lightning-conductor met with opponents in France, and particularly in academic circles, which included the advocates of the Medieval method: to shake the belfries in order to drive the thunderstorms away. Franklin actually sought to demonstrate by statistics that the belfries from which the bells were rung tended to attract lightning. In this absurd dispute between the Ancient and the Modern, the Milanese criminologist and economist Cesare Beccaria, in the name of public interest, sided with his colleague from Philadelphia.

The French and British tradition of the "Lumières" and "Enlightenment" was to be manifested in the publication of encyclopedias. The subsequent editions of *Encyclopedia Britannica*, and particularly the ninth edition published in Edinburgh in 1879, assert a continuity between the science of the 18th and the 19th century. As an imitation of the academic model where the pupil succeeds the master, this progressive vision features as many harmonious stages in the progress of a general theory. The theoretic conflicts between the physicists seem to have been reserved for a later, more feverish and disputed, epoch.

Enlightenment and subsequently Positivism referred to the scientists of electricity as *philosophes de l'expérience* ("experimental philosophers"). This expression, if it is still necessary to repeat it, reflects the supremacy of inductive theory on practice. We are poles apart from modern quackery and its occult powers. Most European countries were to celebrate, in a true "Pantheon of Electricians," some national hero, popular figures whose names became the name of an entity. This was the case in France: Ampère, Arago; in Denmark: Oersted; in Italy: Galvani and Volta; in England: Faraday, Joule, Maxwell; in Germany: Gauss, Helmholz, Ohm. The theoretic discoveries could remain a spiritual satisfaction. Witness Maxwell's electromagnetic demonstration, which around 1868 demonstrated that the speed of electricity was equal to that of light. The practical applications, for example those of electrochemistry, were to encounter the Baudelarian crowds of the World Exhibitions and the metropolitan avenues. The producers of monumental sculptures made "bronze" multiples in galvanoplastic and the gas lamp-posts in cast iron were plated with copper.

Beyond theory we have the history of electricity, an abstract phenomenon explained by mathematics and by the hypotheses on the nature of the material and of light, which was to be shown externally through the physicality of its architecture and through geometric signs, omnipresent in the land: from the city to the mountains, from the home to the desert. This issue of *Rassegna* purports to examine the type of works comprising the architecture of electricity.

1. From C. Collodi, *Puhm! Puhm! Puhm!, comincia l'esame*, 1885. Quotation taken from the *Corriere della Sera*, 28 September 1995.
2. B. Cohen, *Benjamin Franklin's Science* (Cambridge Mass.- London, 1990).
3. G. Chrystal, "Electricity," in *The Encyclopedia Britannica, a Dictionary of Arts, Sciences, and General Literature*, 9th ed. (Edinburgh, 1879).

1. A. Chernikov, architectonic fantasy. Early 1930s.
A. Chernikov Archives, Moscow.

2. Paris World Exhibition, 1937.
Giant insulators, on the right *La Fée Electricité* by Raoul Dufy decorating the pavilion.

1

Sara Protasoni

City Lights: The Visible and the Invisible

"We are entering the deafening realm of electric music. The third movement of the William Tell symphony unexpectedly follows Ludovico's eulogies. Elegant cabinets enclose melodious messes."[1] The oxymorons of Gadda's prose (delivered for the occasion at the electrotechnical pavilion of the Fiera di Milano in 1932) are here taken as a metaphor for the relationship between the built city and the secret passage, within it, of electricity. A relationship that develops on the border between the world of energy production – where chaos, noise and continuous movement show themselves in a complicated intertwining, in a mess – and the world of electrical current, where the tangle is unraveled in the urban network, the noise of power plants becomes the plot of melodic lines and the architectures of the metropolis appear as "elegant cabinets" which "enclose melodious messes."

Two different figures are evoked here by the word "metropolis". On the one hand, the large city, the result of the uncontrollable urban growth under way in the industrialized west after the mid-19th century. The city of congestion, in relation to which hygienists, sociologists, economists, writers, artists and, perhaps a little late, architects and engineers, have set themselves at a critical distance, formulating possible solutions and alternative models.[2] On the other "the Futurist city, similar to an immense tumultuous builder's yard, with all its parts lively, mobile, dynamic, and the Futurist home, similar to a gigantic machine;"[3] or again, the Super-American city of Boccioni and Musil. A large artifice, inside which "air and earth form an anthill, traversed by the various levels of communication routes. Aerial trains, trains on land, underground trains, air-delivered mail, automobile chains darting across horizontally, fast elevators pump masses of men vertically from one floor of traffic to the other; at the junction points you jump from one means of transportation to another, and their pace which pauses between the two racing, roaring speeds, a syncopation, a small gap of twenty seconds, inconsiderately sucks in and swallows the people, who barely have time to exchange a couple of words in haste in the intervals of that universal rhythm."[4] The same image can be found in the background of the description that Simmel traces of the blasé attitude in the context of the big city, where the incredible variety of stimuli and experiences determined a perceptive condition and, at the same time, a method of representation of space and time of the city that was wholly new. Or again, the city, the protagonist and

2

theater of Modernism, inside which a tension is at work between what is fixed and immutable on the scene and what is in endless movement and transformation.[5]

It has been observed on many occasions that urban planning, as an art and/or technique of building cities, and civil engineering,[6] as the technique of designing, realizing, managing urban networks – water, plumbing and sewerage, waste disposal, heating, transportation, communication systems, and, of course, electricity – began to focus on their own disciplinary statute after 1850, in the space that was created between the city as it was and the city as it was meant to be according to intellectuals and technicians. At the service of the city authorities, these followed a "path of normal science," referring the anomalies and dysfunctions brought by urban planning back into known rules and procedures.

In this history, during the course of which the metropolis, or rather, the *Großstadt*, was regarded as a huge work of urban engineering, which was supposed at one and the same time to *function* and *be beautiful*, electricity was to find its own scenic space in the streets, in the world exhibitions and in the large department stores, showing itself as what made lighting and transportation in the city efficient, inexpensive, but above all pleasing; something that, although it could be produced, distributed and sold, referred above all to the sphere of representation.[7] Besides, even if already around 1850, with the construction of the first urban networks for the distribution of gas, lighting had become one of capitalism's fields

3. E. Hénard, *Une Ville de l'Avenir : Vue à Vol d'Aéroplane.*
From: E. Hénard, *Etudes sur les transformations de Paris* (Paris: Editions L'Equerre, 1982).

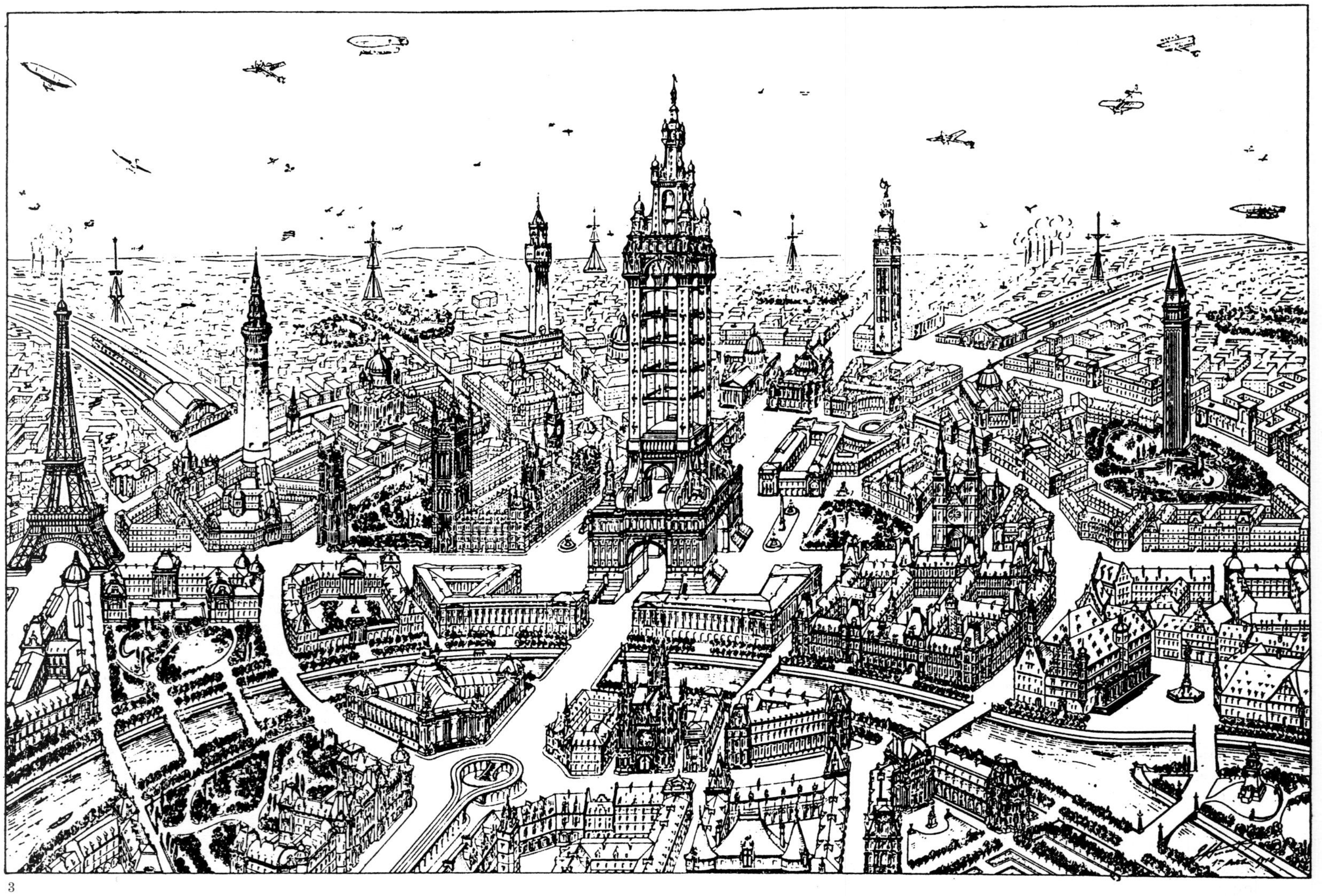

3

4

4. F. Depero, *Grattacieli e tunnel*, 1930.
Museo di Arte Moderna
e Contemporanea di Trento e Rovereto.
© Fortunato Depero by SIAE, 1995.

5. *La circolazione futura e i grattanuvole a Nova York*.
From: *L'Illustrazione Italiana*, 31 August, 1913.

of economic achievement,[8] the electric light takes a further step, highlighting a way of conceiving innovation that presupposes the existence of public service in its modern conception. From this moment on, one of the main tasks of the city authority became that of responding to the economic needs of individual citizens and businesses, not only to make production possible, but also so that this might contribute to accelerating the circulation of capital. The city and/or state companies that were formed in the large European and American towns were to address an indeterminate number of users – with their boundaries continually expanding – and would have at the same time to guarantee the efficiency and good value of the management of the service when faced with times and ways of use that would tend to be concentrated in certain periods and to stagnate in others.[9] The public technical equipment of the city was to be considered and designed to also represent this "potential" wealth and efficiency, to show a technical culture in which technological innovation and the pleasure of experience were bonded together.

The architects devoted themselves to composing sections of roads which, though offering themselves as pure technical-functional solutions, take on the typical configurations of the *passage*. "The *Passages* shone in the Paris of the Empire like enchanted grottoes [...]. The sudden explosion of electric light cancelled out the virgin paleness of these passages, which suddenly became difficult to find; they exercised a black magic of doors, as though they were looking from blind windows inside themselves."[10]

In the main conferences and urban-planning exhibitions of the turn of the century, the road section is highlighted as a theme of urban composition, not only as the design of detail, which describes, together with façades and trees and the shape of the streetlamps, the arrangement of the mains under the paving, but also because it refers back to an image of the city underground as a continuous space, with respect to which what has been defined as "underground urban planning" takes shape.[11] At its origins lies a way of looking at the city and its territory in terms of "systems" (parks, fortifications, roads, or, more exactly, mobility) which had already been depicted in the projects by Haussmann in Paris and, through the re-elaboration work by Hénard[12] and others, was to be at the center of the international debate on urban planning of the early part of the century. Following shortly after, the technical networks (water, electricity, gas, etc.) were also to join

5

6. Battersea power station, London, 1925-33. Phase A.
From: Fondation Electricité de France/IFA, *Architecture de l'électricité* (Paris: Norma, 1992).

6

7

7. Advertising for Selenium.

8

8. A. Chernikov, architectonic fantasies, 1928-31.
From: *Avantgarde II, 1924-1937. Sowjetische Architektur* (Stuttgart: Gerd Hatje, 1993).

the systems making it possible to represent the modern city, which lent itself to a configuration through abstract ideograms, meant to represent the space contested between schemes of operation and models of representativeness.[13] In *Mechanization Takes Command*, Sigfried Giedion placed research into movement in space – and in particular into methods of representing this – at the origin of the mechanization of production, which in the modern age becomes the responsibility of a specific professional figure: the production engineer.[14] A similar path seems to have been followed by Modernistic urban planning, committed throughout the 20th century to giving shape to increasingly abstract schemes.

We said that the metropolis also shows itself as a way of perceiving and representing urban space and time, applying itself to recording a tension between what is fixed and immutable and what is continually transforming. In the city it could be said that one's gaze attempts to fix an element which exists above all – as electricity does – in the dimension of virtuality, thanks to that which it makes possible.

The same gaze can be traced back to the basis of Boccioni's compositional research and his reflection on the landscape of the metropolis, as it appears in the manifesto *Architettura Futurista* of 1914, the same year in which Sant'Elia and Chiattone exhibited their designs in Milan. In this treatment the continuous reference to the figure of the spiral is fundamental. "The architectural environment of the city is transformed into a winding direction. We live in a spiral of architectural forces. Till yesterday building revolved in a successive panoramic direction [...]. Today we begin to have an architectural environment which develops in all directions: from the lit-up basements of large department stores, from the various tunnel levels of the metropolitan railways, to the huge heights of US skyscrapers."[15] It is the discovery that as many realities can exist as there are points of view.

We must here refer to the results of the most recent reflections on modernity, and in particular, regarding the experience of space and time, to the work of David Harvey and Stephen Kern;[16] but we also need to bear in mind those studies which were applied to the reconstruction of the affirmation of a new theory of sensitivity between the late 19th and early 20th centuries, when progress in the theoretical-scientific fields made it necessary to rethink the organizational principles of perception which were at the basis of the different methods of representing the world. This was a full-blown scientific revolution which obviously exercised considerable influence over artists, writers and architects. Moving from a new theory of sensitivity – one might say from a new aesthetics – architects now began to formulate the problem of city aesthetics (*civic art* or *art urbaine*). It is worth recalling certain facts: Sitte had published *The Art of Building the City* in Vienna in 1889; in 1903, *Esthétique des Villes* by Charles Buls, published in Brussels in 1888, was translated into German; finally, from 1898 on the various international conferences of *L'Art Public* were held, in Brussels in 1898 and 1910, in Paris in 1900, in Liège in 1905. At those venues the debate on the "beautiful city" was related back to the analytical awareness of the forms of the cities of the past, to the search for geometric rules for the reconstruction of the urban scene, which had been lost in the new metropolitan dimension. With respect to this picture, the essay that August Endell published in 1908, with the significant title of *Beauty of the Metropolis*, marked a position that clearly broke with the past: the metropolis "is ugly." Yet in its ugliness, it "appears as a symbol, as the strongest expression of a *Kultur* that has moved away from nature, from simplicity, from ingenuousness."[17] To the gaze of the architect this reveals a "hidden beauty which does not speak to the senses, which is accessible to those who enter the world of work with the power of their imagination." In a "world full of energy and movement," beauty cannot be other than the result of an intellectual exercise which, moving from vision, is capable of going beyond this to construct landscapes. It is a beauty that is located beyond the visible.

Electricity and Urban Form: Strategies of Concealment

Thomas Alva Edison presented his patent in Europe for an urban lighting system during the first international exhibition of electricity, which was inaugurated in Paris in August 1881; on that occasion, not only the new carbon filament electric bulbs were exhibited, but also the largest dynamo to be built up to that time, capable of supplying a full-blown distribution network. From 1882, numerous companies, including the Società Generale Italiana di Elettricità Sistema Edison and the German Deutsche Edison Gesellschaft für angewandte Elektrizität, which was to become the Allgemeine

9

9. Paderno-Milan plants. View of the electric station.
From: Società Generale Edison di Elettricità in Milano, *Impianti Paderno-Milano*, n.d.

10. Paderno-Milan plants. Planimetry of Paderno d'Adda plant.
From: *ibid.*

Elektrizität-Gesellschaft (AEG), worked to introduce the new system in Europe too.

The "commercial" success of Edison's patent highlights the fact that a large city such as New York, where technological concentration (which is also physical concentration) makes the "field transfer" of scientific and technological conquests more immediate, be it in its own right capable of amplifying enormously the perspectives of application of each new discovery.[18] In relation to this, the role played by the periodical and general press must not be underestimated; the associations of professionals in the sector that form in the main US and European centers express themselves through these: not only because they make it possible to delineate, with respect to a discipline – electrical engineering – that was still in the construction phase, the symbolic barriers that should separate scholars and non-scholars, but also because of the influence they have over the construction of a "technological common sense," increasing the expectations and the relative consumption relating to the new lifestyle that electricity allowed individual citizens. Yet in both cases these were themes that it is only possible to hint at here.

The Italian way to electrification highlights another important element of connection between the progress of electrical engineering and the urban phenomenon: the urban form seems to be capable of directing research towards possible systems of energy distribution.

In an article published in 1901 in the magazine *L'Elettricità*, we read: "the special configuration of the city of Milan and the activity of city life, with everything grouped there around a single center, offered quite favorable conditions for the development of an electrical distribution network of the Edison type. In Rome, however, the distribution of electrical energy posed a very different problem. On the one hand, there were several centers of activity, separated by very large distances; on the other, large areas intermingled with developed areas, where it would not be possible, or at least it would be useless or very ineffective, to create an electrical distribution."[19] Therefore in Rome the distribution network could not propose Edison's "central" model again, but had to be structured into more or less innervated parts, to support the non-homogeneous nature of the built-up areas. Above all, it had to cover considerable distances between production stations (energy was still generated by coal-powered dynamos) and the points of supply, all this limiting the size of the installation and, above all, the running costs.

In the alternating current system by Gaulard used for the first time by the Società Anglo-Romana for the lighting of Rome in 1886 (it must be said that the experimental nature of this initiative was increased by the nearly total ignorance of the possible consequences of such a high difference of potential on the component parts of the installation), a high-potential underground network was envisaged to supply a large number of transforming substations, each of which was at the center of a low-potential network. The main plant was situated in an off-center position with respect to the shape of the city; the network, initially stretching for 6 km, was soon extended to 19. Research into the possibilities of improving transportation over long distances, while keeping costs down, suggested by the particular orographic conditions of the city and its shape, in 1892 led to the creation of a hydro-electric power plant in Tivoli, which made use of the waterfalls near Villa d'Este, and of the first experiment in the transportation of energy (in this case two-phase) over long distances: the Tivoli-Rome line.

In 1882 the Società Generale Italiana di Elettricità Sistema Edison was formed in Milan; its founder, Giuseppe Colombo, in direct contact with Edison, had managed to obtain a franchise for Italy – after long negotiations with the French Compagnie Continentale Edison, who had guaranteed themselves the exclusive right to use on the European market. In 1883, a year after the construction in New York of the world's first electric power plant, the new plant (the first in Europe) was inaugurated in Milan. It was a single generating plant, situated near Piazza del Duomo, supplying an underground network that stretched across an area with a radius of approximately 500 m.[20]

The Edison system made provision for the use of direct current. Users joined the network through derivations in series; by increasing the amplitude of the system and the number of derivations, the furthest points from the generating station (the outlying locations) were therefore penalized, unless conductors with enormous sections or very high voltages were installed, and these it was nevertheless very hard to use for civil purposes. The image of the Milan plan as the section of a tree, with which in 1885 Cesare Beruto illustrated the new city planning scheme, presupposed an analogous network model.[21] The reflection could also be extended to the subsequent phase of electrification of the Milan area, where the passage to the hydro-electric model was delineated: it is on that occasion that the first network in Europe for the transportation over long distance of three-phase energy was realized. The Paderno d'Adda plant, near Milan, completed in 1898 as a complex system of production, transformation, transportation and distribution of energy on a regional scale, which probably not only constitutes the most advanced synthesis – for that period – of scientific theories and technical innovations in the field of electrical engineering to be developed during the course of the 19th century, but also forcefully proposes once again the way of looking at the relationship between the city and

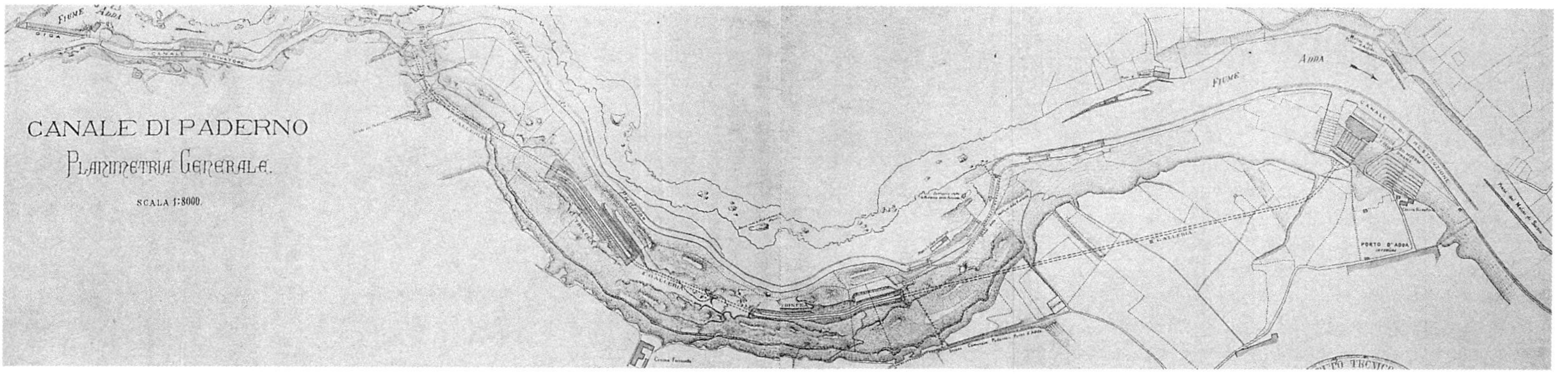

10

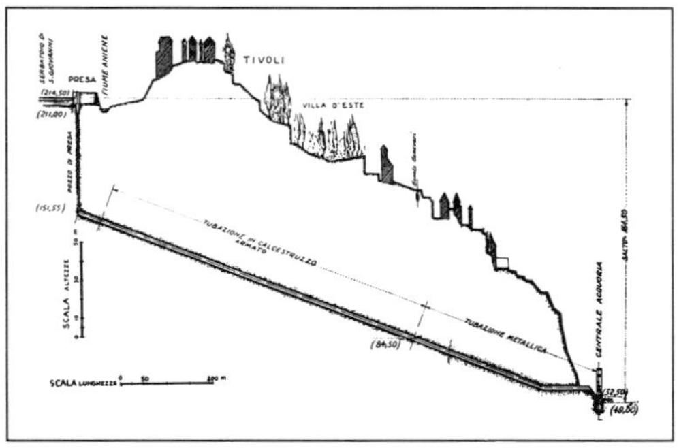

11

11. Pressure water pipe of the Tivoli plant (Aniene).
From: *Nel cinquantenario della Società Edison: 1884-1934* cit., vol. I.

12. Power trasmission Tivoli-Rome (1890-2). The hydroelectric station below the ruins of Maecenas' villa.
From: Associazione fra Esercenti Imprese Elettriche in Italia, *Notizie sui principali impianti elettrici d'Italia* (Milan, 1911).

its territory that had been refined by Lombard polytechnic culture, from Carlo Cattaneo on: the landscape as an "immense monument to the constructive arts" and technical design as a manifestation "of civilization and progress."[22]

The direct current in the original Edison system, and in any case electricity in general in the early years of its use in the city, is still considered on a par with a fluid or a combustible gas, which must be transported by special pipelines to the point of use: a monodirectional flow that is structured through derivations in series. Besides, the early public lighting systems not only imitated the intensity and the quality of gaslight, but even took up its distributional and administering model. The quality of the supply, entrusted to private initiative, was a function of the distance from the places of production, scattered around the city; the disadvantages of the system were attributed to the losses of energy that occurred during the course of transmission. Faced with these difficulties, innovation was initially sought so as to limit to the utmost the need for transportation: it was therefore referred above all to the points of supply and to their displaced positions. The optimum condition to be achieved seems to be that whereby the production and consumption sites were unified to the utmost.

As regards urban lighting, the paradigm takes the form of a central beacon that illuminates large surface areas from disproportionate heights (like a sun), as in the project by Sébillot and Jules Bourdais, "Colonne-Soleil, Projet de Phare électrique de 360 mètres de hauteur destiné à éclairer tout Paris. Construction monumentale," presented at the competition for the realization of an emblem for the World Exhibition in Paris in 1889. The same idea can already be identified, on a smaller scale, in the Edison light tower placed at the center of the electricity pavilion at the Columbian Exhibition in Chicago in 1893, and had been realized in certain US cities in the second half of the 1880s. Behind these proposals lies the image of a territory shown, in the night, through juxtaposition of enormous illuminated circles. Electricity is superimposed over it like a sort of grid, recording the differences between places by measuring the distance from the luminous nuclei, and its rhythm marked by a natural temporality, where day and night follow one another.

At this point it is necessary to make an observation. The radio signal sent in 1912 throughout the world from the Eiffel tower, and also the immense signal towers that

12

Hénard, in 1910, imagines arranged in a crown formation around the center of his "future city" (which "could serve at the same time as wireless telegraph stations; and it is not impossible to foresee the instant exchange of communications between all the countries of the world with the help of these large poles of the celestial sphere"),[23] reveal a very different conceptualization. At the basis there is the idea of a space that can be cancelled out at the moment when world communication eliminates any distance: a space that can suddenly be concentrated in a single point. To take up the image of the light on the terrestrial globe, a space irradiated homogeneously by constant light, inside which the variation in intensity must be the outcome of a project, of a plan.

This is a fundamental passage to understand the relationship that, at the start of the 20th century, links the consolidating of a practice of urban designing such as "civil architecture" with the delineating of urban planning (as a new urban science) and of the *Großstadt* (as its paradigm). The most complete formulation of the paradigm of the *Großstadt*, even if it almost certainly arrives in second place, after Wagner's, in 1910, is by Hilberseimer (1927). "We can only oppose theoretical plans with a demonstrative character against the chaos of today's metropolis. Their task is, starting from current requirements and operating at a purely abstract level, to develop the basic principles of urban planning, so as to obtain general rules that would make it possible to solve certain concrete problems."[24] Expressed here is an antagonism between the theory and the rules of urban development and its built reality. The metropolis exists in the dimension of the diagram, fundamentally as a means of representation. The architecture of the city identifies its themes of planning and only then, locally, does the space rediscover its own dimensions and its own physical form. This approach brings the problem of infrastructure networks in the modern city back to the relationship between their traces (mains and rails) and the ground. The architect's task will be to set the conditions that must regulate *the artistic configuration of the road layout*.

In 1903 the essay by Charles Buls on city aesthetics was translated into Italian and the author was invited to Rome that same year to hold a lecture on the subject "Aesthetics of the Street." In an article published in 1910 in the magazine *L'Elettricità*, Buls' conference was quoted literally: "for the large avenues, for the

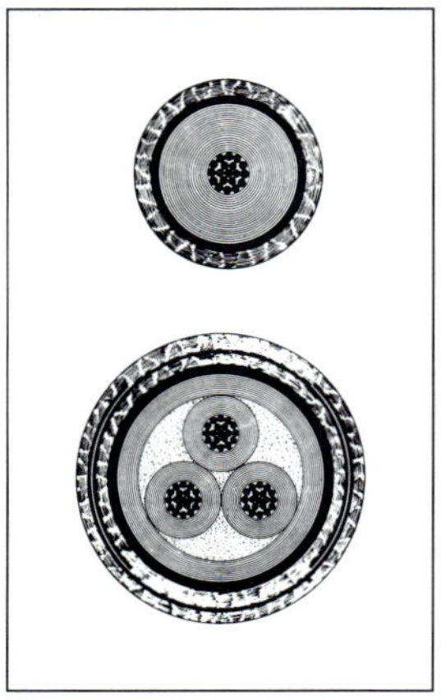

13

13. Section of electric cables. From: *Nel cinquantenario della Società Edison: 1884-1934* cit., vol. I.

14. Rome, electric underground cables and secondary stations for public distribution of electricity (1886-97). Archivio Storico ENEL, Rome.

15. Milan, the distribution net in 1898. From: *Nel cinquantenario della Società Edison: 1884-1934* cit., vol. IV.

roads and picturesque squares, then, we need to ban the system of aerial cables without remission, and hide the electric streetcar equipment under the ground, as is done in Brussels."[25] The separation between the road level and the infrastructure level and the concealment of networks respond to the parameters fixed by the whole community with respect to city aesthetics, in Paris as in Vienna. But it is above all the city of Vienna that at the start of the century, through the work of Karl Lueger, makes a considerable effort in the field of public law to promote its own image in Europe that would be effective both on the representational level and on the operational one; in the notebooks compiled by Lueger, the secret geometry of the gas, water and electricity networks was revealed.[26]

The Invisible City and the Distribution of Sensible Reality to the Home

Yet, that same *Panorama* of Vienna that for Karl Lueger embodied a classical ideal of synthesis between "usefulness" and "beauty", in Musil's vision became a transitory image, if not even illusory, of a reality that cannot be wholly grasped; "an enchanted grotto," exactly like the *passages* in the Paris of the Empire according to Benjamin.[27] Vienna "like all metropolises, consisted of irregularities, alternations, precipitations, intermittencies, collisions of things and events, and, in the midst of all this, points of abyssal silence; of rails and of virgin lands, of a great rhythmic beat and of the eternal discord and confusion of all rhythms; and all in all it was similar to a seething blister placed in a materialized recipient of houses, laws, regulations and historical traditions."[28] A dimension of perception seems to be lost in this description: that from which derives the sense of duration and which makes memorizing possible. Now, against the background of the modern experience of the space and time of the city, the presence is perceived of "a materialized recipient of houses, laws, regulations and historical traditions" without distinction of types. To talk of a city of stone different from the city of men seems to have lost meaning. As much as we can see, Wagner as well stops placing the city that is being built as the exclusive center of his own work. Indeed, in 1910, he writes: "the way of life imposed by our time will arouse many other things that we can barely imagine today;"[29] further beyond, the urban project is also expressed as a desire for a flexible space capable of supporting technical progress, the future. According

14

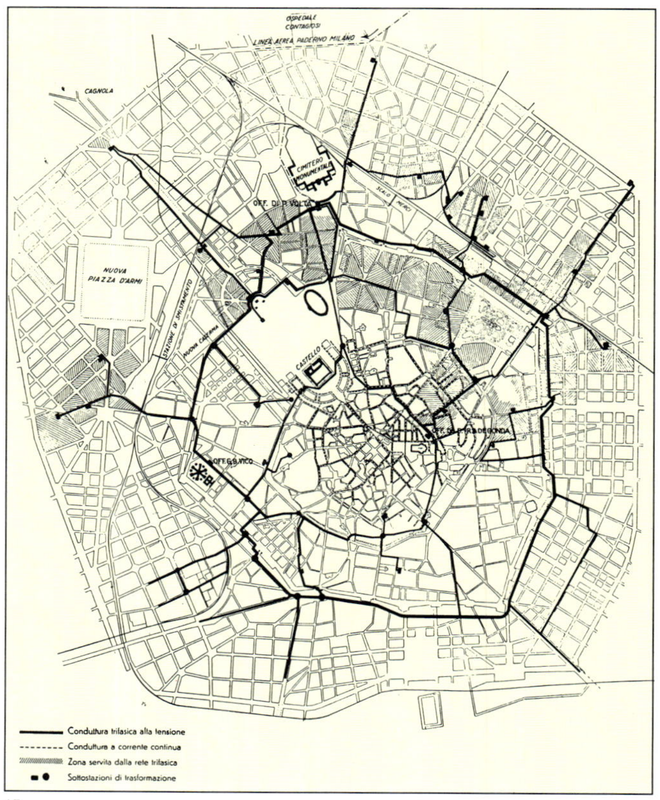

15

to Wagner, architecture must represent its correspondence to the requirements of life and of modern sensitivity, placing simplicity and uniformity among its own fundamental compositional principles, going increasingly towards a reduction of complexity. The effort which, in the same years, the electrotechnical engineers began to make to represent in a single ideogram quantities and times of consumption, supply costs, output available relative to phenomena in which the variation is continuous and not uniform, goes in the same direction. Innovation at all costs, the creation of new systems, was no longer pursued. For the engineering of the urban networks, the economic management and supervision of the patrimony began to constitute a privileged field of intervention in the city.

Since the first attempts, the electrification of the city was aimed at an ideal condition where the dispersions due to the transportation of energy were reduced to zero; the extreme condition was the one whereby there was a virtual relationship between the place of production and the place of use, which was realized – instantaneously and without any need for conveyance – only if and when necessary. This result was partially achieved in the field of communication, in a sector within which limited power was used and it became possible, therefore, to contain costs of plant and operation. The network was then able to move close to the condition of a tool of homogenization of space, also capable of suppressing the differences between city and countryside, between core and suburbs.

Among the alternatives that the planners developed with respect to the problems brought by congestion, this model of network finds numerous applications: the most obvious reference is that by Wells and his *Anticipations of the Reaction of Mechanical and Scientific Progress Upon Human Life and Thought*, of 1902, in which he hypothesizes an urbanized region extending limitlessly and innervated by roads, railways, telegraph lines, telephones, pipelines for the pneumatic mail.[30] Even the research carried out by the "Soviet Disurbanists" between the 1920s and 1930s (such as the "Green City Moscow" proposal developed by Ginzburg and Barsh in 1929) were connected explicitly to the campaigns for electrification promoted by Stalin. Finally, Wright's "Broadacre City" (1932) represents a profoundly anti-urban model where the physical dispersion of the built areas and the realization of a lifestyle totally centered on the individual are made possible by a hypothetical system of land roads and areas crossed by strictly private means.

16

16. Advertising poster for Incandescenza Auer. Archivio delle Stampe Achille Bertarelli, Milan.

The theme of the relationship between the public and private spheres, due to the relationship with spatial models that tended towards homogenization, was not greatly debated by the architects of the start of the century. At the time when it became possible to live everywhere according to a model of behavior that made use of technology – but we could even say of electricity – to guarantee oneself *happiness*, the representation of progress can no longer only be referred back to the characteristic forms of urban space, to the traditional sites of the social. Conversely, the risk becomes one of falling into theatrical fiction. The relationship between Coney Island in New York, which Rem Koolhaas highlighted in *Delirious New York*, is paradigmatic from this standpoint.[31] And once again, referring to a place where a "technology of the fantastic" that almost exclusively makes use of light and movement to produce its effects, the theme of electricity returns. But what must be underlined here is that, when faced with the possibility of reaching, through technological investment, a condition of indifference in relation to the position with respect to traditionally central locations, Modernism necessarily locates private domestic space at the center of the reflection on the form/function relationship.

The utopian horizon with respect to which the spatial research of the Modern Movement takes shape is not therefore only represented by the great theoretical models of Le Corbusier and Hilberseimer; it is also that traced around the start of the 1930s by Paul Valéry: "the distribution of sensible reality to the home." A reality which, deprived of matter, shape, dimension and duration, becomes energy capable of producing vital experiences anywhere and at any time – experiences of space, of time, of ourselves and others, of the possibilities and dangers of life.

"Like water, gas, electrical current reach our homes from afar to meet our needs with almost zero effort, thus we will be fed by visual or auditive images, which will appear and disappear with a minimal gesture, almost with a nod. As we are accustomed, if not subject, to receiving energy in various forms into our homes, so we will find it very easy to obtain or receive those rapid variations or oscillations with which the organs of our senses, which perceive and integrate them, make everything we know. I do not know if a philosopher ever dreamed of a society for the distribution of sensible reality to the home."[32]

1. C.E. Gadda, "A zonzo per la fiera," in C.E. Gadda *Azoto e altri scritti di divulgazione scientifica* (Milan: Scheiwiller, 1986), p. 93.
2. The main models of urban expansion perfected between the end of the 19th and the start of the 20th century, as different authors have already highlighted, entrusted to technological innovation in the sector of transportation and communication the task of resolving the problems linked to congestion. See, among the other examples, E. Howard with his *Tomorrow: A Peaceful Path to Social Reform* (1898), upon which he was to construct the extremely influential movement of the garden city, with the pilot experiments of Letchworth (1903) and Welwyn (1919), and his extreme reinterpretation in "Broadacre City" (1932) of F.L. Wright; or the "Ciudad Lineal" (1882) by Arturo Soria y Mata and the proposals of the "Soviet Disurbanists" in the 1920s and 1930s.
3. A. Sant'Elia, "L'architettura futurista," (Milan, 11 July 1914), published in *Lacerba* (1 August 1914), now in P. Hulten (ed.), *Futurismo & Futurismi*, catalogue of the exhibition at Palazzo Grassi (Milan: Bompiani, 1986), p. 419.
4. R. Musil, *L'uomo senza qualità* (Turin: Einaudi, 1957), pp. 26-27.
5. On these themes, see, among others, D. Harvey, *La crisi della modernità* (Milan: Saggiatore, 1993), in particular pp. 33-42 and 322-46.
6. Cf., among others, A. Abriani, "Dal sifone alla città," in *Casabella*, No. 542-543, 1988, pp. 24-9, and A. Guillerme, "Sottosuolo e costruzione della città," *ivi*, pp. 30-5; W. König, "Massenproduktion und Technikkonsum. Entwicklungslinien und Triebkräfte der Technik zwischen 1880 und 1914," in *Propyläen Technikgeschichte. Netzwerke Stahl und Ström* (Frankfurt am Main-Berlin: Ullstein, 1990), pp. 265-552; A. Guillerme, "From Lime to Cement: the Industrial Revolution in Civil Engineering, 1770-1850", in *History and Technology*, 1986, pp. 5-67.
7. It could be asserted, with Benjamin, that roads, like "the world exhibitions were the schools in which the masses excluded from consumerism learnt the empathy with the value of exchange. *Look at everything, touch nothing.*" Cf. W. Benjamin, *Parigi capitale del XIX secolo. I "Passages" di Parigi*, ed. by R. Tiedemann (Turin: Einaudi, 1986), p. 264.
8. Cf. W. Schivelbusch, *Luce. Storia dell'illuminazione artificiale nel secolo XIX* (Rome: Pratiche, 1994).
9. The reconstruction of the debate that developed in Italy after the approval of the Giolitti law of 1903 on the "direct taking-on of public services by Municipalities," related back by the journals of the electronic sector to the problem of the organization of city and state organizations for the production and supply of electrical energy, would from this standpoint be a very interesting field of investigation. Cf. D. Calabi, "I servizi tecnici a rete e la questione della municipalizzazione nelle città italiane (1880-1910)," in P. Morachiello and G. Teyssot (eds.), *Le macchine imperfette. Architettura, programma, istituzioni nel XIX secolo. Atti del convegno. Venezia ottobre 1977* (Rome: Officine, 1980), pp. 293-316.
10. W. Benjamin, *Parigi capitale del XIX secolo* cit., p. 727.
11. Cf. A. Abriani, "Dal sifone alla città" cit., p. 28.
12. Particularly in his works at the Musée Social from 1908 to 1913. Cf. E. Hénard, *Etudes sur les Transformations de Paris* (Paris: Librairies-Imprimeries réunies, 1903-1909), and "Die Stadtmitte von Paris und die Knotenpunkte des Verkehrs," in *Der Städtbau*, No. 10, 1910, pp. 109-13.
13. As regards the insertion of the technological networks among the elements of the urban phenomenon by urban planning, the seminars on urban planning held in Berlin between 1907 and 1915 by J. Brix and F. Genzmer are very significant. Cf. *Städtebauliche Vorträge aus dem Seminar für Städtbau an der königlichen Technischen Hochschule zu Berlin* (Berlin: Wilhelm Ernst & Sohn, 1907-1915).
14. Cf. S. Giedion, *Mechanization Takes Command. A Contribution to Anonymous History* (Oxford, 1948).
15. U. Boccioni, *Architettura futurista*, 1914, now in Z. Birolli, *Umberto Boccioni. Altri inediti e apparati critici* (Milan: Feltrinelli, 1972). Written by Boccioni between the end of 1913 and the start of 1914, the manifesto was for a long time attributed to Marinetti.
16. Cf. D. Harvey, *La crisi della modernità* cit., and S. Kern, *Il tempo e lo spazio. La percezione del mondo tra Otto e Novecento* (Bologna: Mulino, 1988).
17. A. Endell, *Die Schönheit der großen Stadt* (Stuttgart: Strecker und Schröder, 1908).
18. The spatial reflections of the industrial inventions and innovations in the electrical field have been studied, among others, by A.R. Pred, *Sviluppo industriale e sviluppo urbano negli Stati Uniti* (Milan: F. Angeli, 1978); S. von Weiher, *Berlins Weg zur Elektropolis* (Berlin: Stapp, 1974); T.P. Hughes, "The City as Creator and Creation," in J.P. Kleihues (ed.), *Berlin-New York: Like and Unlike. Essays on Architecture and Art from 1870 to the Present* (New York: Rizzoli International, 1993), pp. 13-31; cf. also J.A. Tarr, M. Rose, J.W. Konvit, "Technological Networks and the American City: Some Historiographical Notes," in *Flux*, No. 1, 1990, pp. 85-91. Those who, among others, have investigated the social consequences of electrification, are: T.H. Hughes, *Networks of Power. The Electrification of Western Societies* (Baltimore: Johns Hopkins University Press, 1983); A. Laszlo and I. Masulli, "Elettricità e vita sociale," in G. Mori (ed.), *Storia dell'industria elettrica in Italia. Le origini. 1882-1914*, vol. 1 (Bari: Laterza, 1992), pp. 645-96. As regards the construction of technological common sense, cf., on the American example, C. Marvin, *Quando le vecchie tecnologie erano nuove. Elettricità e comunicazione a fine Ottocento* (Turin: Utet, 1994), and, on the history of lighting, W. Schivelbusch, *Luce* cit.
19. G. Mengarini, "Impianti elettrici italiani," in *L'Elettricità*, 20 April 1901, p. 248.
20. Cf. G. Colombo, "Illuminazione elettrica," in *Milano tecnica dal 1859 al 1884*, ed. by the College of Architects and Engineers of Milan (Milan: Hoepli, 1885), pp. 459-73.
21. Cf. C. Beruto, *Progetto del Piano Regolatore della Città di Milano. Relazione all'Onorevole Giunta Municipale* (Milan: Tipografia Bernardoni e Rebeschini, 1885).
22. C. Cattaneo, "Su la densità della popolazione in Lombardia e su la sua relazione alle opere pubbliche," vol. 1, file 1, January 1839, pp. 29-52. Cf. O. Selvafolta, "Progetto e cultura politecnica: la prova dell'uso," in P. Caputo (ed.), *Milano percorsi del progetto*, pp. 257-306.
23. E. Hénard, "Le città del futuro," report made to the International Conference of Town Planning in London in 1910, now in D. Calabi and M. Folin (eds.), *Eugène Hénard. Alle origini dell'urbanistica: la costruzione della metropoli* (Padua: Marsilio, 1972), p. 192.
24. *L'architettura della grande città di Ludwig Hilberseimer* (Naples: Clean, 1981), p. 13.
25. G. Aurini, "Sistemazione estetica degli appoggi e dei castelli telefonici," in *L'Elettricità*, 28 July 1910, p. 56.
26. Cf. K. Lueger, *Die Wasserversorgung sowie die Anlagen der städtichen Elektrizitätswerke, die Wienflussregulierung, die Hauptsammelkanäle, die Stadtbahn und die Regulierung des Donaukanals in Wien* (Vienna, 1901); K. Leuger, *Die Erbauung des Wiener städtischen Gaswerkes* (Vienna, 1906); but above all *Ausstellung der Stadt Wien*, catalogue for the Milan Exhibition in 1906 (Vienna: Gerlach & Wiedling, 1906).
27. W. Benjamin, *Parigi capitale del XIX secolo* cit., p. 727.
28. R. Musil, *L'uomo senza qualità* cit., p. 6.
29. O. Wagner, *Die Großstadt. Eine Studie über diese* (Vienna: Anton Schroll, 1911), now in O. Wagner, *Architettura moderna e altri scritti* (Bologna: Zanichelli, 1980), p. 118.
30. H.G. Wells, *Anticipations of the Reaction of Mechanical and Scientific Progress Upon Human Life and Thought* (London: Chapman and Hall, 1902).
31. Cf. R. Koolhaas, *Delirious New York. A Retroactive Manifesto for Manhattan* (Rotterdam: 010 Publishers, 1994), pp. 29-79.
32. P. Valéry, "La conquista dell'ubiquità," in *Scritti sull'arte* (Milan: Guanda, 1984); published for the first time in *De la musique avant toute chose* (Paris: Editions du Tambourinaire, 1928). The same phrase, restricted to the first part, was quoted by W. Benjamin in *The Work of Art in the Era of Technical Reproducibility*.

17. El Lissitzky, *Coureurs*, 1926 ca.
Galerie Berinson, Berlin.

17

17

1

Hervé Guillemain

Everyday Miracles. Domestic Electricity in France Between the Two Wars

In its oscillations between utopian scientism and superstition, between Verne and Huysmans, the end of the 19th century shuddered at the sight of the magic show offered by a mysterious and inaccessible fairy: electricity. The fascination was caused by the classical mechanism of attraction-repulsion. Public opinion, submerged by futuristic images, aware of the quasi-divine power of an immaterial form of energy, considered the possibility of electric apocalypses, gigantic explosions, endless blackouts. This fear was merely the hidden face of a new freedom: creating light, electricity facilitates transportation and communications, thus bringing ancestral solitudes to an end. It gave rise to a period during which rapidity and dynamism became key words. Morand and Lenin praised the energy of the century – "religion of the 20th century" according to the former, "pillar of communism" to the latter – a promise of freedom from darkness and immobility, and thus from Nature.

Before 1920 the applications of electricity remained the appanage of the community – the streets were lit – or of industry, which the electric engine stimulated to expand. Despite the periodic waves of "electromania" electricity, apart from its uses for lighting, remained far from the private dimension, and the single individual essentially remained hostile. In 1914 the illustrated *Larousse* classified it under the voices describing the engine, light and the telephone.

But the electrical revolution does not lie in the innovation concerning the new light source, something which had already been achieved with the use of petroleum and gas, but rather in the revolution of the instantaneousness which unifies the world and, as McLuhan puts it, "enables any remote village to become a center."[1] The image was overturned in the period between the two wars: the electric mystery was replaced by the work of electric household appliances. Electricity therefore was no longer a dream, but the tool which determined the realization of dreams. Daniel Rops, doubting the modern triumphant mechanization, witnessed this evolution: "Electricity will, to an increasing extent, be the sole mechanic force truly able to produce, because it enables us to replace, in all sectors, human work with the automatic work of the machine [...] which performs exactly the same role as slaves in antiquity."[2] The transformation from a fairy into a slave symbolizes the transition of electricity from a spectacular form to a source of energy for domestic uses. This evolution solves the new problem created by the feeling that History and its progress was accelerated, and that a rhetoric of the "vital" time was becoming diffused. Henri de Graffigny, author of manuals on the domestic use of electricity, as early as 1928 predicted: "The rhythm of life has been accelerated, and it is no longer possible to dedicate the same number of hours to domestic work as in the past, because time has become a precious asset."[3] However, the increasingly predominant functionalistic image of electricity did not dispel that of the fairy. In 1941 Paul Devaux, in his *Histoire de l'électricité*, asserted: "That electricity is a faithful servant is proven every day by a thousand humble miracles [...] In the early years of the century our fathers, dazzled by the marvels of Science, gave us the now outdated name 'Electric Fairy'. If this expression makes us laugh today, it is because we are ungrateful."[4] The image of electric divinity adorned by purifying virtues, "immortal," "Indian princess whom one cannot touch without dying," continued to exist. In 1954 Francis Ponge, in his *Texte sur l'électricité*, once more compares the two images: "It is a matter of a maid with the character of a princess."[5]

2

3

At the World Exhibition of 1937 electricity was presented in two forms, that of light and that of the machine, the fairy and the servant. As Raoul Dufy's work *La Fée électricité* [today at the Musée National d'Art Moderne of Paris] shows, the Compagnie Parisienne d'Electricité (CPDE), which commissioned the painting and the pavilion where it was exhibited, chose to unite the dream and the everyday reality, thus gratifying Paul Léon's wish to create, in the exhibition for which he was responsible, an alliance between "Homo Sapiens and Homo Faber, an indispensable union between the spirit and the hand."[6] The fresco serves as a background for the powerful turbines installed in the rooms, while the images of one of Paul Grimault's first cartoons, *Phénomènes électriques*, shot by means of procedures that were revolutionary in France, such as the "hypergonar" objective and technicolor, was projected on a panoramic screen outside. In the movie – which recounts the adventures of Mister Plus and Mister Minus in the world of electricity – the fairy-servant shows all its technological power, at the same time conserving the typical fascination of a magician's show.

A gap seems to exist between the image, sublimated by the requirements of the Exhibition, and the reality of the extent of electrical consumption which was still limited in a country, France, which played the role of poor

4

On the preceding pages:
1. R. Dufy, *La Fée électricité*, 1937. Lauros-Giraudon.
2. J. d'Ylen, image of the "Electric Fairy" in 1920. Bibliothèque Forney, Paris.
3. A.M. Cassandre, advertising for Als-Thom, Electricity at the service of the electric housekeeping, 1931. Roger-Viollet, Paris.

4. *Les dix commandements de l'électricité*, 1925. Compendium for new generations. Photothèque EDF, Paris.

5. In favor of electric housekeeping: Electrolux advertising in the 1930s. Bibliothèque Forney, Paris.
6. The domestic taylorization of middle-class homes. From: *Art Ménagers*, 1933. Photothèque EDF, Paris.

relative in comparison with the United States or Switzerland in this field. In spite of the diffusion of household appliances after 1931, the domestic quota of the electricity consumption had not yet become a strong element in the French economy. The new energy remained a luxury of the middle classes, due to a prejudice that remained deeply rooted in spite of the efforts of the Laval government and the alignment, since 1930, of the rates to those of gas, although the costs of the apparatuses remained three times as high.

But what were the causes of the French delay? In 1932 90% of the population already received electricity, and the distribution network could indeed be considered as practically completed. Moreover, the production was sufficient to meet everyday needs, since it had been quadrupled in the period 1914-30, and had not suffered the consequences of the great crisis. While the apparatuses were neither perfect nor acceptable aesthetically speaking (in most cases one continued to clothe the electric modernity in traditional attires) they had been long since invented (1900-20) and marketed (1920-30). The French delay therefore seems to be a consequence of the insufficient urbanization of a country inclined to conserve its rurality, since electricity remained, at the time, a decidedly urban phenomenon. The conservatism of the countryside delayed the diffusion of the new form of energy, and in addition to obstacles of an economical nature it was therefore necessary to solve the much more difficult one represented by mentality. In order that the use of electricity in the homes could become widespread, it was necessary to work simultaneously in three areas: to break down the bourgeois image of this energy, sustain its social usefulness and make it decidedly inexpensive.

In order to recover the French delay the propaganda of CPDE, the forerunner of EDF, aimed at educating a reticent and badly informed population on the problem of the introduction of new technology in the private sphere. A "Maison de l'Electricité" was created in boulevard Haussmann, to demonstrate the domestic uses to the general public and as a service center. The façade featured the following text: "Don't do what electricity can do for you." The "Centres Ménagers" opened as of 1928 were much more organized and efficient; they gave courses on how to use the new electrical apparatuses, a basic education aimed at transforming apparently revolutionary gestures in everyday automatisms. It was above all necessary to prove that French cooking did

5

6

not become less refined when prepared with the help of electrical appliances, considered expensive and unable of producing savory dishes; to promote their diffusion – which was also encouraged by specific credit facilitations and a lease service – the centers therefore appealed to the great French cooks. But the education projects also had to focus on the future generations of consumers. The CPDE therefore trained a teaching staff which could teach domestic education in the schools. The children were an ideal target for advertising campaigns aimed at the schools, featuring bonus-points decorated with images inspired by the domestic use of electricity, and electric appliances offered as prizes at the end of the school year. One also resorted to games (the "electric lottery" as of 1930) and to the movie theaters where shorts on the subject were shown during intermissions.

As of 1928 the CPDE also published, with success, "electric diaries" with practical advise, as well as a *Bulletin d'Information* distributed to professionals in the electricity sector, architects and medical doctors. Finally, vast operations were planned in order to draw attention to the introduction of electricity in the homes: in 1937 the experiment of the "electric village," based on the free distribution of electric appliances to the inhabitants, was launched in Saint-Hilaire-sur-Helpe, in the north of the country. Aimed at achieving a wide diffusion of the energy in a rural context, this project was anything but convincing, and electricity was to remain a typically urban phenomenon for a long time.

The private corporations continued their advertising campaigns incessantly in order to increase the need for electricity. As the "Lux-fascination" and "Lux-révélation" campaigns witness, the rhetoric of the "electric miracle" was associated to that of the modern slave. The following message can be found in one of these advertisements: "I, the home valet, knock at your door. I am Electro-Lux, I bring comfort and security." Based on a number of key words, these slogans were presented as attractive images. Hygiene, in its purifying dimension (cleanliness, illumination) capable of healing the sick and making hair beautiful, turns electricity into a clean energy in comparison to the nauseating petroleum and gas. Electric heating is recommended to obtain "a perfect equilibrium of the lungs, the mucous membranes and the skin."

Safety associated with the absence of flames is underscored in the same way as the low cost: the slogan "Agricultural workers are increas-

7

7. Advertising of the Association pour la Promotion de l'Electricité, 1930s. Photothèque EDF, Paris.

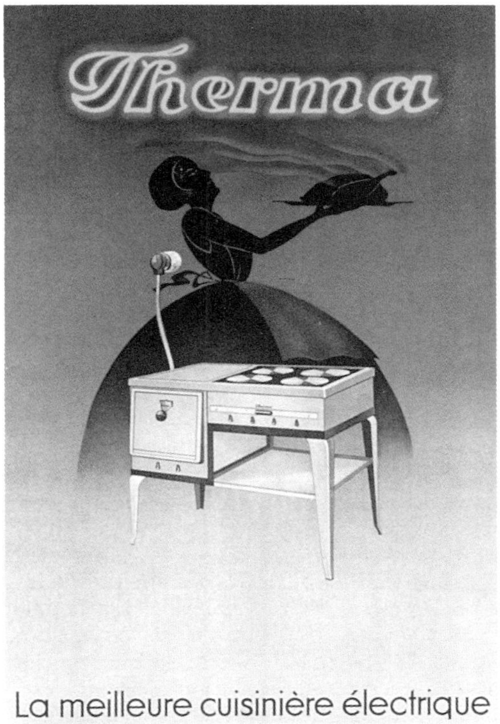

8

8. Advertising on the advantages of the electric kitchen, 1935. Bibliothèque Forney, Paris.

9

9. The first signs of the Electric house. Inside view of a Paris house in 1925. Roger-Viollet, Paris.

ingly hard to find and costly, replace them with electricity" was displayed along the roads in the agricultural regions. The speed, or "the miracle of the switch" seemed to appeal to the collective imagery at least as much as the idea of comfort: unpleasant tasks become a game. The movements are simple and are done with a smile, while the key phrase "At your service" resounds in the mind. One vaunts "modernity," repeating that "the ancient gestures of domestic life are never abandoned," a slogan borrowed from American models. Thus, in 1938, an advertisement of the Société Electrique de Paris et du Rhône claims that it guarantees "comfort in the home." "Domestic life goes on without noises, shocks, dust, smells and in the reassuring and warm intimacy reserved only for the family members."[7]

Despite these efforts the dream of "everything electric" had not yet been realized in the late 1930s. The specter of the electric home continued to hover, nurtured since the beginning of the century by Robida's novels, by the catalogues (*La Maison électrique* by Knap of 1913) and by some avant-garde experiments, such as that of the "familistery" of Guise, equipped with own electric ovens. And in 1924, in his article "Que sera la vie domestique future," A. Christofleau describes a phantasmagoric social future, revolutionized by the presence of household appliances: "The hearth, understood in an individual or familiar sense, will tend to disappear [...] The automation of community services will suppress most of the typical activities of a retired and individualistic existence, in favor of a new life style, no longer social but group-oriented [...], no longer domestic but based on a system of buttons, light indicators and microphones."[8] He speaks of "modern phalanstery" and "skyscrapers," thus offering a surprising syncretism of the modern utopias. Apart from making the electric dream topical, the realization of the electric home produced its first fruits in the 1930s. Lighting, heating and cooking were the most common functions, but the most diffused apparatuses, those which meet the requirements of making the private space functional, were the small household appliances. The first washing machines (1923), fearfully complicated, and washing appliances, with their countless faucets, discouraged the families due to their costs and difficult use. With the exception of the radio, only the iron and the vacuum cleaner invaded the homes on a large scale. Improbable and today forgotten products attracted the clients: heating plates, anti-cough ozonizers, electric egg-cookers, immersion water heaters. This inventory, worthy of Boris Vian, witnesses the imaginary revolution experienced by a generation that identified the housewife with the worker in *Modern Times*: the machine must serve to save time and to achieve a rationalization and an efficient management of space. These criteria of domestic Taylorism are concretized in a kind of futurism-style automatism such as, for example, in the electric groom praised by the catalogues in 1928. The phantasm of electricity, far from being materialized between the domestic walls, commits the social-political protagonists to a reflection on possible innovative solutions. One of the consequences is the publication, in 1935, of the *Notes sur l'électricité à l'Eglise*, aimed at helping the parsons to bring about a true electrical revolution: is it possible to use organs, microphones and electric bells without modifying the perception of the prayers on the part of the believers?

Rather than a technical problem, the matter of the domestic use of electricity was a social problem. When World War I had just finished bathing the country in blood, the conservative public opinion (especially the League of Mothers and Housewifes) did not look favorably on a phenomenon which would impede the feminine labor used during the war from

10

10. Anemostat advertising.
Photothèque EDF, Paris.

11. The wide range of electric household appliances, Lutra advertising, 1930.
Bibliothèque Forney, Paris.

12. The child, target of "electric" advertising. The Calor heat, 1930.
Photothèque EDF, Paris.

returning to the hearth. The danger therefore consisted in seeing the neo-Malthusians prevail, since electricity would certainly have favored the reduction of the consistency (the implosion) of the family nucleus. The battle against the decreasing birth rate and the moral decadence induced by domestic comfort once more triggered, with certain figures of paternalists in the circles of social Catholicism, the eternal diatribe against the machine, symbol of modernity.

On the other side of the political arena, inspired by Hugo and Lenin, one tended to magnify a form of energy that was considered a basis of future social democracy. The anti-republican far left sought, in electricity, a tool for suppressing the alienation of domestic work, while the Republican regime saw it – as witnessed by the World Exhibition of 1937 – as a guarantee of progress and a weapon against obscurantism; electricity, in fact, guaranteeing a comfortable domestic life, could make everyone a member of the middle classes.

Jules Louis Breton, an engineer, revolutionary socialist (1890) and then Republican Minister of Public Hygiene (1920), in 1923 promoted the experiment of the Salon des Arts Ménagers, aimed at applying science to everyday life. With this event (and the magazine *Arts Ménagers*) whose objectives were educational rather than commercial, above all Breton hoped to "democratize" electricity, diffusing it in all homes. As to the conservative values and their contrast with those of the Republic, women – at the center of the debate – recommended another approach. As of the 19th century the image of electricity has been associated with that of the woman; its creative power associates it to the figure of the mother, its extraordinary elements to that of the witch and its representation to that of a charming, desirable woman. After 1920 the press and advertising no longer aimed their messages at men, mere experts on technical notions, but rather to women, the consumers. Their ideal client was an elegant, modern woman belonging to the middle classes, whom this saving energy had freed, while her child, with an idyllic family environment in the background, gestures to attract her attention. The great idea of feminism in the 1920s – to Taylorize housework and make the figure of the housewife professional – was imported in France by Paulette Bernège (through her magazine, *Mon chez-moi*, then *La Femme au foyer*, and the League of domestic organization founded in 1925). Her efforts, influenced by the American example, were aimed at making housework a science in which each movement was codified. In her most important work, *De la méthode ménagère* (1928) electricity is assigned a symbolic position within the domestic life of the 1930s. The mechanization of the housework could thus free the woman or, on the contrary, enslave her. On the other hand, it is difficult to define the political orientation of this movement. While it is true that Paulette Bernège does not spare herself with regard to Republican slogans – "Our republic has not abandoned the age of slavery. We complain of the shortage of domestic servants. But the issue is badly posed. We seek human beings as servants, while, in our period of science and democracy, the sole servants are of a mechanical kind" (1929); "Every woman can become bourgeois [...] this would be a revolution accomplished without fire or flames, but with electricity" (*Mon chez-moi*, May 1927) – it is nevertheless also true that her ideas concerning the "Woman as dispenser of well-being for the family" were approved by the traditionalism of Louis Marin. Also some industrialists supported her initiatives, which often resembled a kind of family lobbyism in favor of electricity, as her professional reconversion as a dealer of household appliances witnesses.

11

12

This debate, interrupted by the war, was however only in its early stages. The consumers' society promoted by the glorious 1930s was to resume when the electric home had become a reality.

1. H.M. McLuhan, *Pour comprendre les média* (Paris: Seuil, 1968), p. 56.
2. D. Rops, "La jeunesse et l'ère du machinisme," in *La Revue des deux mondes*, January 1928.
3. H. de Graffigny, *Le Trésor du foyer*, 1928, p. 93.
4. P. Devaux, *Histoire de l'électricité* (Paris: PUF, 1941), p. 5.
5. F. Ponge, "Texte sur l'électricité," in *Le Grand recueil*, vol. I (Paris: Lyres, Gallimard, 1961), p. 96.
6. P. Léon, *Du Palais royal au Palais Bourbon, souvenirs*, 1947.
7. *Je sais tout*, October 1924.
8. A. Christofleau, "Que sera la vie domestique future", *Les dernières nouveautés de la Science* (Paris: Hachette, 1925).

13. Vitry Sud power station. Cover of *L'Illustration*, 15 March 1941. From: Fondation Electricité de France/ IFA, *Architecture de l'électricité* (Paris: Norma, 1992).

13

1

1. L. Figini, G. Pollini, the Electric House at the 4th "International Triennial Exhibition of Modern Decorative and Industrial Arts," Monza, 1930. Night view.

2. P. Bottoni, Electric House: dining room, kitchen, pantry and servant's room with the movement of dishes. Plan. Milan Politecnico, Archivio P. Bottoni.

3. Electric House: the hall between the corner pillar in Eternit pipe and the double-glass conservatory-window. On the board, advertising of the firms sponsoring the Electric House.

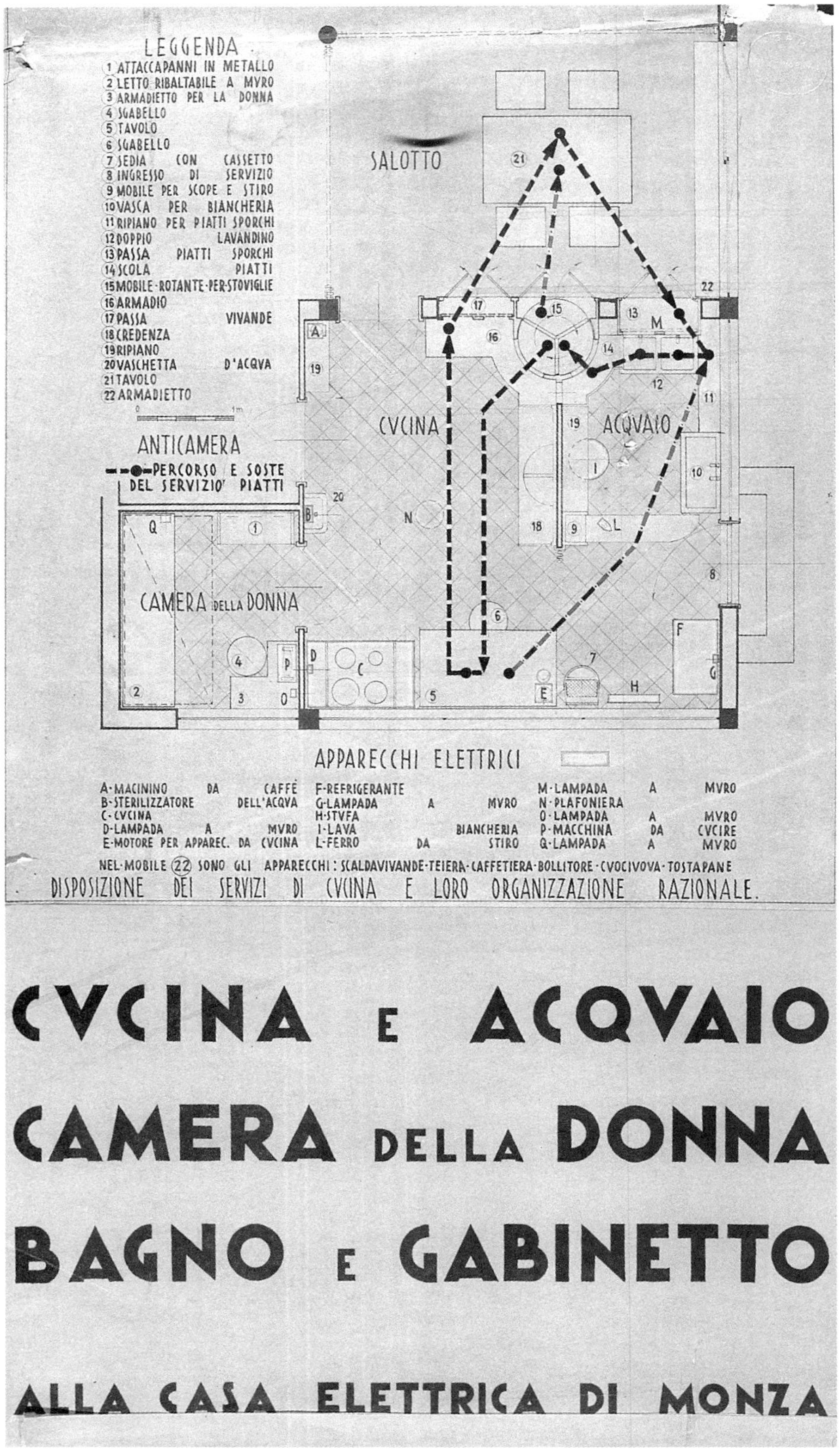

2

Giacomo Polin

The Electric House

Designed during the first months of 1930 by Figini and Pollini, with the assistance of Bottoni for the bathroom and kitchen areas and Frette and Libera for part of the furniture, the Electric House was inaugurated in May that year in the park of the Villa Reale of Monza, within the context of the 4th "International Triennial Exhibition of Modern Decorative and Industrial Arts."

Along with it, a mutual design effort of some of the "architects of the Italian rationalistic movement" created as setting and application of electric appliances, another two home-pavilions were presented: The Holiday House by Gio Ponti and Emilio Lancia, and the Casa del Dopolavorista by Luisa Lovarini, an architect from Bologna.

The three small houses purported to represent as many "models" which were the consequence of the economical contribution of the respective customers: La Rinascente and De Angeli Frua, manufacturers of materials for interiors, funded the country house designed by Ponti and Lancia, the Opera Nazionale Dopolavoro the Casa del Dopolavorista, and the Edison Company the Electric House, designed by the young members of the Gruppo 7.

The house, conceived as a very modern and at the same time luxurious experiment, based on an initiative of Ponti elaborated by the young rationalists Figini and Pollini, was designed in the form of a villa due to the ambiguous and therefore "open" character of this building type, where the Triennale and Edison joined forces in a rational experiment with a life-size setting.

The Electric House, as all the other "practical experiments" staged during previous Biennales, had to demonstrate the concrete application of the more or less structural relationship established between the architects and the different manufacturing contexts. Moreover, the Electric House was based on a data that was intrinsic to the Monza shows and the applied design approach they pursued: its "commercial" objective or promotional scope represented a phenomenon assumed by the designers as a material to be formalized with the greatest responsibility.

After the exhibition modalities were completely separated from the commercial aspect, the Electric House purported to solve the problems posed on the application level by converging phenomena in terms of materials and communicational ideology. The extraordinary "life size" exhibition of the devices and appliances inside a home that was aware of its role to the point of assuming the most appropriate

3

4

4. Electric House: the kitchen with the pass-through cupboard by P. Bottoni. Small electric household appliances everywhere.

5. P. Bottoni, the kitchen of the Electric House. Perspective view. Politecnico di Milano, Archivio P. Bottoni.

6. Double-glass conservatory-window in the Electric House.

7. Electric House: living room with stair to the first floor. Floor covered with linoleum, pillar in Eternit pipe, hardware in chromium-plated aluminum, armchair and table by Adalberto Libera. Electric household appliances everywhere.

5

6

7

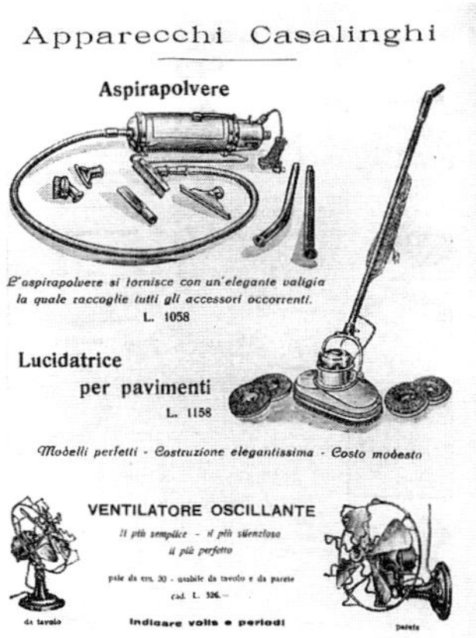

8

8-10. Floodlights, irons, warmers, hair-driers and other electric household appliances on show in the Electric House.
Edison Catalogue, 1930.

name, the Electric House, established the concrete image of that which, with intelligent conciseness, had been defined "the magic transition of art-industry-business-public."

The pavilion which was, at the same time, a country house, a laboratory of life styles, a set design, a "typical" setting for applied technology and finally a home, the Electric House, represented the complete accomplishment of this "magic transition" and its most accentuated result.

This effort of didactic experiment and concrete research of a complete, modern and efficient home model, was realized through the characteristic device which from the very start had been typical of the Biennales and the Triennales: "interior design as a way to furnish." In it, and at a realistic, even life-size scale, the fundamental objective of the exhibition was manifested; its exasperatedly didactic character as well as its conscious personality. As in the "realistic settings" which had characterized the exhibition methods of the Biennales and Triennales already from the start, thus also in the Electric House – where the purpose of every fragment, material and household appliance was to express its own contextual situation in the current functional reality – it was the "need for architecture" which determined the interior arrangement: but this time the design really reflected the complexity of the architecture. The Triennale exhibition included, in addition to "decorative goods," also the structures functional to their presence, through the construction of houses: the "architectural superattraction" whose necessity Fortunato Depero had championed, and the application of design to the problem of the "aesthetics of sales" represented the generative nucleus of the operation of the Electric House.

In September 1929 Pollini had written an article titled "L'elettricità e la casa," electricity and the home, for the magazine *Natura*; he here described and commented the realization of his project for an Electric Apartment in Bolzano, which anticipated, within the context of a regional trade fair, the design content of the Electric House.

During 1930, in the pages of *Natura* – whose stated objective was to study the most interesting aspects of the natural sciences, but which did not disdain to recount the vicissitudes of deforestation and thus the construction of hydroelectric dams in the mountains, the physical-geographic aspects of the process of production and distribution of electric energy, even providing practical advise for "rational lighting" – one developed the subject of electricity as a natural-artificial phenomenon, and that of its "applicability" to architecture.

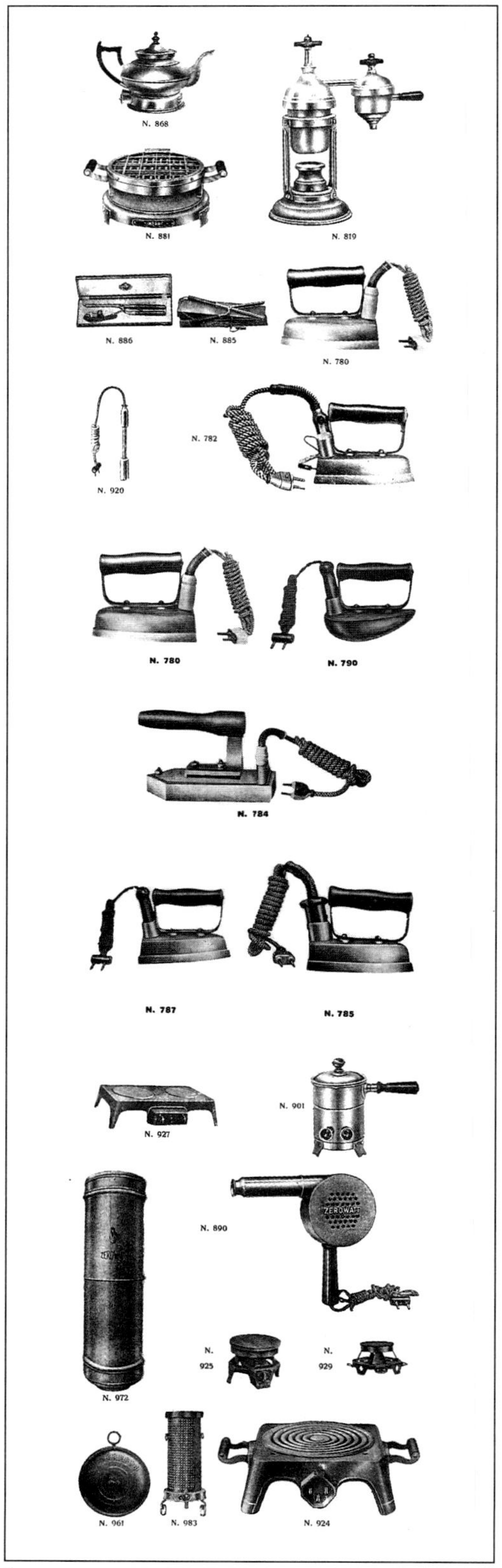

9

In the meanwhile Piero Bottoni devoted himself, parallelly to the others, to the observation of the same necessities and, by no chance with respect to the subsequent participation in the project of the Electric House, wrote two articles for the magazine *Illuminotecnica* in which he expressed his adhesion to the technological theme. As of January 1929 Bottoni had published, month after month, on the cover of *Illuminotecnica*, a specific drawing titled, coherently with the image presenting the crossed beams of very strong reflectors as background for an "extremely rational" modern city, "Luci nella notte," lights in the night; the metaphor of a design commitment bound to the contingent technological opportunities (like electricity) could not have been more explicit. The two articles, published by the magazine which more than any other – and not only by virtue of its name – stressed the importance of the lighting-technology relationship of light, and its effects on architecture, are titled respectively "Illuminazione dell'architettura" and "Architettura dell'illuminazione" and contain, for an analysis of the intellectual itinerary of Bottoni, the author of the "electric kitchen," many interesting motives.

Generally speaking, an "electric debate" was triggered: this debate was mainly instigated by the Edison company, which was founded in 1884 and which had become, around 1930, an industrial group of considerable size and solid branching, operating in the field of the production and distribution of electric energy. Through its magazine *Parole della Edison ai Suoi Utenti*, as well as through the official organ of UNFIEL, *L'Energia Elettrica*, Edison announced its industrial and social projects, which were to transform not only the makeup of a single manufacturing sector, but the very forms of work and everyday life: the electrification of the homes. This goal, which saw an exemplary opportunity of advertising in the construction of the Electrical House, was first and foremost pursued through a campaign that tended to favor and popularize electrical lighting as a modern technology and form: "rational lighting," thus defined evidently in homage to a climate that was by then diffused and perceptible by the masses, became a goal at which to aim on a short term basis for an entire, powerful industrial group.

To further clarify the genealogy of the project: in the Autumn of 1929 Pollini realized in Bolzano, as mentioned above, the Electric Apartment which purported, on a reduced scale but with the same pretext, to "make the scope and destination of electric apparatuses more evident by exhibiting them in the environments where they will actually be used in the home." Commissioned by The Azienda Elettrica Consorziale and designed by Pollini shortly before his definitive move to Milan, this Electric Apartment contained, in a nutshell, the same characterization of "rationalistic realism" as the subsequent and much more comprehensive Electric House, enabling its author to assert, in the article of September 1929 which had inaugurated the collaboration of the Gruppo 7 members in *Natura*, "as modern forms of aesthetics are the spontaneous result of a rational solution of the problems posed by present-day life." In fact, it was a matter of an as yet undefined project, in which the problem of the exhibition of the electrical appliances was solved by means of indirect artificial light and the use of mass-produced furniture; but the primary importance attributed to electricity as an intangible organizer of the space made Pollini's limited project assume a didactic clarity. The design concept of a realistic and at the same time abstract interior, which was to be accomplished in the Electric House, was subject to a first, significant outline in the few square meters of the Bolzano apartment.

In 1929 the estimated figure of installed and functioning electric household appliances was about one million five hundred thousand (the majority of them were small or very small, as boilers and irons): the electrification of the home, as a rational project, required aggressive promotional initiatives in order to catch up the accumulated delay, and to convince growing masses of Italians of its necessity. From the initial incentive, in the rational illumination project, to exploit electrical energy as a simple and rational source of light, one proceeded with another, and potentially much more totalizing project aimed at the general use of household appliances, which the architects, and even more so if rationalists, had to be able to further with their work.

On the plan of the Electric House the designers thus traced the dotted line of the itinerary that the visitors were to follow, starting at the entrance and visiting all the rooms on the ground floor in a process of circular development that ended at the back exit.

The scheme for a "rational visit," the conception of a possible rational use as early as at the design stage, thus represented an impulse in favor of the sole possible form of "consumption" of the Electric House; the latter, which could in theory be consumed but which was, in fact, only consumed in the course of three months, could actually only be "passed through," since the intangible and continuously suggested sensation of a ready-to-use reality materialized in a static inactivity. Although it was real, the Electric House did not work, it was not used. By explicitly inviting the visitors to explore it, it lived for consumption but was never consumed.

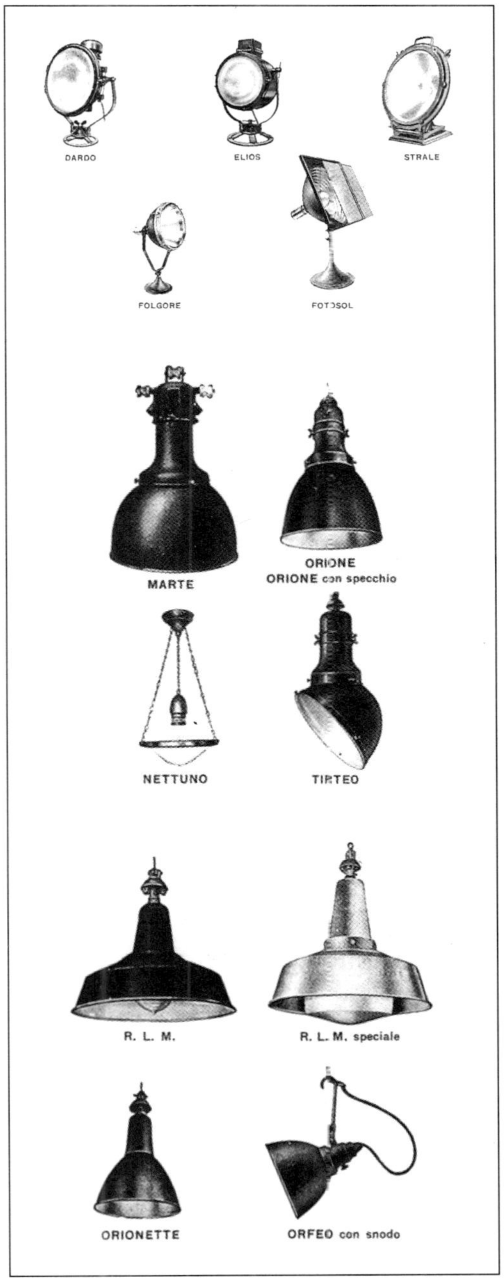

10

Figini and Pollini, to a certain extent going beyond the strictly pragmatic indications provided by the historical moment and by their own intention in terms of design scope, purported to overcome, in the Electric House, the narrow architectural limits of the exhibition pavilion, instead preferring to design a villa, or rather a "villa-pavilion." This gave rise to the basic ambiguity of the Electric House, the essential reason why it represented an aggressive attempt to mediate, as an open experiment and at the same time as a compromise.

The design of a villa-pavilion, whose distinguishing trait of formal creativity and openness towards the environment, characteristic of the villa as a building type, therefore integrated and mitigated the rigorously technical and didactic content of the exhibited arrangement of an industrial production.

This camouflage was part of the exhibition plan, since the various functional areas of the Electric House featured a considerable number of technical applications to the problem of electricity, a vast range of samples of lamps and household appliances.

In the rooms of the Electric House the different fragments of the entire domestic production had to be arranged so as to summarize their comprehensive functionality by integrating them in a whole. The requirement of a "re-education on architecture" pursued by the Triennale was also accomplished in the precise functional exhibition of the apparatuses and the furniture, which expressed the modern ambitions of the building organism on a small scale. What the small objects, scattered everywhere, mainly served to show was the "economic value of the renewal of taste," the transformation, to a large extent structural, of the objectives and forms of the products for the home.

The participation of industry in the design of a setting for technology in the home (and in the nature) therefore required that the designers made an effort of realism that was so perfectly defined and disseminated that it became abstract, above and beyond the concrete possibilities of domestic furnishing on the part of the (limited) mass for which the Electric House was to serve as an example. Around the living room other works of "pure art" were scattered in an attempt to counterpoint the presence of the electrical appliances: a sculpture by Arturo Martini, two etchings by Carrà, a bas relief by Fausto Melotti represented a "setting" for the entire arrangement, seeking to make the contemporary exhibition of ozonizers, plate-warmers, floor polishers, fans and teapots, all of them obviously electric, more harmonious.

Hugues Fiblec

When the Cathedrals Were Electric

Paris "Ville Lumière"

Around 1900 the myth of the "Electric Fairy," moulded as representation of a political direction, namely Republican liberalism which, with its constant appeal to the virtues of "progress for everyone," sought an authoritative cover for an enterprise of strong ideological connotations, appeared in France. The bourgeois century of steam, tallow and coal was to be followed by the new socialist era of electricity, hygiene and modern comforts. Lenin's famous aphorism, according to which socialism would merely be the sum of "Soviet and electricity," found its omens in Emile Zola who in 1901 prophesied: "The day will come when electricity will be diffused throughout the world [...] It will circulate in the cities almost as if it were the very blood of social life."[1]

This does not take into account the "rival" form of energy of the period, gas. Indeed gas, the most democratic of light sources, had soon become associated with the very concept of progress and the republican regime. And it is precisely the republic which ungratefully inflicted the first spectacular blow to gas in 1887, when the Opéra Garnier, already present during the empire, saw to replace its eight thousand gas outlets with as many incandescent bulbs. While it is true that the promoters of gas and electricity continued to rub shoulders in the competition until the 1930s, it is equally true that the public at large had already made its choice; indeed, the World Exhibitions of 1889 and 1900 consecrated the triumph of electric lighting. Thus, in 1900, the Electric Fairy found its enchanted exhibition palaces and saw the erection of its "cathedral-power stations." The gas industry had been relegated to the city outskirts, while the electric one succeeded in penetrating to the interior of the city. This kind of invasion paradoxically indicates that the techniques were still in their embryonic stages; in fact, the electric current could not be transported over long distances without considerable dispersions and it was consequently necessary to install the stations in the heart of the area where it was to be consumed.

This shows the entity, in the form of the architectural integration of the production station in the townscape, of the challenge faced by the industrialists. A dual challenge, in that it was on the one hand necessary to reassure the Parisians who feared the harmful consequences of these malodorous and noisy plants and, on the other, to convince the most hostile opponents of the usefulness of the new source of energy. Thus, in the competition between gas and electricity, the architecture of the power station immediately revealed itself as a fundamental vehicle for illustrating the advantages and modernity of the "invisible fluid." But faith in scientific and technological progress was not sufficient to make the population accept the negative consequences that the introduction of electricity entailed. Embellished by divine, if not economical, attributes, the Electric Fairy possessed the terrible charm of a Victorious Venus. Is not electricity perhaps, as Paul Morand reminds us, the religion of the 20th century? It was to find its new Suger in architect Paul Friesé, who elaborated a fundamental reflection on the most appropriate architectural expression for this modern equivalent to the Medieval cathedral.

1

Paul Friesé

It is necessary to consider for a moment the singular figure of Paul Friesé (1851-1917).[2] At the time an architect's dedication to industry represented a novelty: to design factories was regarded as a minor activity by the academic orthodoxy, and was efficiently discredited by the monopoly exercised by the engineers in the field. The period spent at the Ecole des Beaux-Arts (1871-8) was as important to him as his collaboration, over a period of five years, with engineer Jules Denfer, whom he succeeded as director of the firm in 1891. Friesé therefore incarnates the typical architect of the "industrial age," also promoted by Eugène Viollet-le-Duc, who had to join together the "art" of the architect and the "science" of the engineer. But Paul Friesé is more than just an industrial designer: his proposals concern both the structure and the form of the building, as well as the organization of the area of production.

It is natural to wonder, first of all, what made this architect-engineer venture beyond the beaten track of industrial architecture, since the power station, in the last years of the century, represented an until then unknown category and thus had to be invented entirely. Friesé, who was born in Strasburg and fled to Paris after the defeat of Sedan in 1870, soon associated with a circle of other Alsatian exiles which vaunted great names of the industrial dynasties of the period. Close to Freemason circles, he supported the idea of a French Alsace, along with numerous Alsatian friends, some of whom were "protesting" deputies at the Berlin Reichstag, but without neglecting to plead the cause of a Europe *with* Germany. His admiration, veined with bitterness, for the German industrial power was to be crucial for the battle led by Friesé, who never ceased to

1. P. Friesé, electric power station of the Compagnie Parisienne de l'Air Comprimé, Paris, 1895-8. Period photo. Institut Français d'Architecture, Paris (IFA).

2. P. Friesé, electric traction station of the Paris Subway at Bercy, 1898-1904. Front on Rue de Bercy. Caisse Nationale des Monuments Historiques et des Sites.

3. P. Friesé, electric traction station of the Paris Subway at Bercy, 1898-1904. Front on Rue de Bercy.

assert the French industrial supremacy, while avoiding the nationalistic temptations of a historicistic architecture. An examination of the situation of the electric distribution in Paris confirms the role as investors from the Alsace-Lorraine region. Renouncing to undertake the electrification of the capital in 1889 the Municipality of Paris entrusted the exploitation of the distribution networks to six licensee companies.[3] These private electricity companies merged, in 1908, into a single company, the Compagnie Parisienne de Distribution d'Electricité (CPDE), which retained the monopoly of the production and distribution of electricity in Paris until it was nationalized in 1946. As of 1900 powerful financial corporations were founded, such as the tandem Empain et Schneider, which controlled, more or less directly, almost the entire Parisian electricity production. This French-Belgian consortium, in association with the bank Bénard, in 1898 ventured on the exploitation of the Paris Subway, the new electric-traction railway. The group's architect – who but Paul Friesé – was entrusted with planning the entirety of the electric infrastructures of two of the six licensee companies, CPDE and the Metropolitan Subway.

A Reasoned Eclecticism

Some may suppose that the power station, according to Friesé, would adopt the same architectural codexes as the spinning mills, the breweries and mills of northern France, true castles of industry. But this is not the case. Electricity is a conquered, tamed type of energy. The architecture of electricity was therefore to be sober, if not absent. Neither fortress nor palace worthy of "thousand and one night," the electric cathedral conceived by Friesé vaunts, as opposed to the formal excesses of the latter, a purist eclecticism, which could be defined functionalistic. This anti-conformistic architect merged the two French 19th-century traditions of constructive rationalism, of both Henri Labrouste and Eugène Viollet-le-Duc. This is where the origins of Friesé's style can be found: first in the rationalistic field and then in that of eclecticism.

The forms of the eclectic culture have served as inspiration for Friesé's first important power station, the generator of the Sector of Place Clichy (1889-91), where he resorted to a classic register in order to integrate the young electricity company in the residential district of Batignolles.[4] Friesé chose to build the office building on the road in order to conceal the big metal hall of the power station; only the tall

2

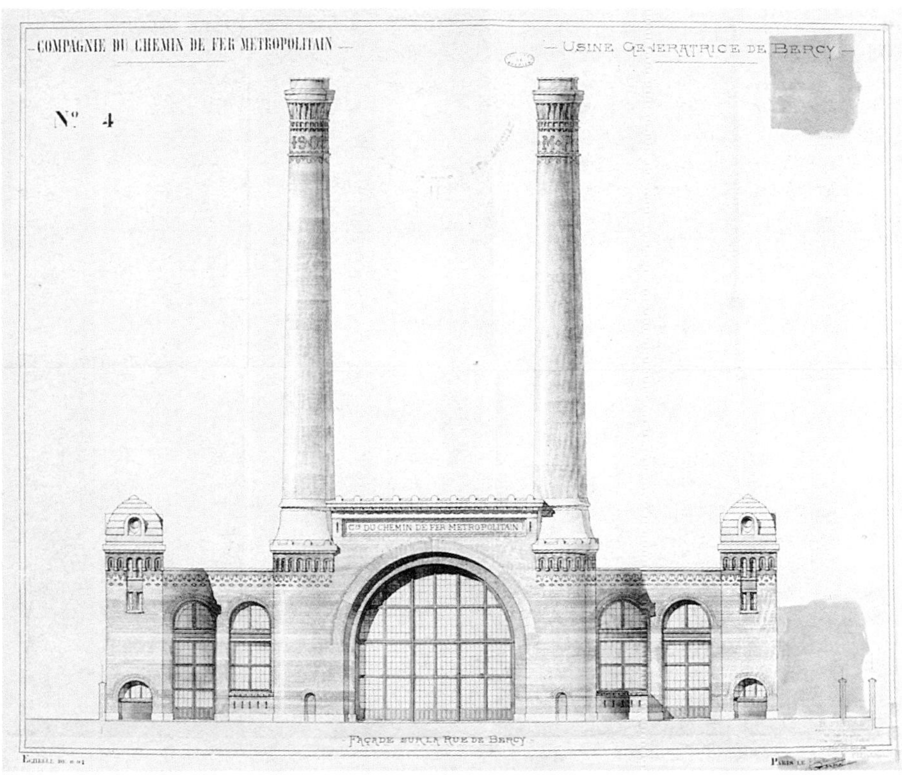

3

4. P. Friesé, electric sub-station
"Opéra", Paris, 1903-4.
RATP.

smokestack rising behind the triangular pediment betrays the building's industrial function. The architectural effect of the façade in brick and stone supported by a metal skeleton was obtained by a mere assertion of the materials: steel girders as architraves, painted bricks, ceramic parapets and a terra-cotta façade. The latter, a true masterpiece of decoration, produced in England, is reminiscent of the Arts-and-Crafts fashion of Edwardian architecture.[5] Friesé soon radicalized his architectural approach in the second, giant power station built in an industrial area by the Saint-Martin canal: the station of the Compagnie Parisienne de l'Air Comprimé along the Quai de Jemmapes (1895-8).[6] A grandiose expression of constructive rationalism, it only apparently deviates from tradition.

Due to the narrow spatial organization of the production stages it was necessary to divide the original plan, which reminds of that of private palaces from the classical period: the bodies of the main building and the projecting symmetric wings house steam engines and electric generators, the structures at the entrance house the offices and the corner pavilions the accumulators, the enormous batteries used to regulate the energy distribution. The internal "distribution" is even more surprising. The shortage of space made Friesé choose a vertical organization of the production area. The architect thus designed an arrangement on several floors, superimposing the boiler room on that where the electric machines were installed.[7] The coal silos and water tanks were installed in the attic. The side façades are an external expression of the distinction of the internal functions: the passer-by, looking from the quai, can easily imagine the vertical division of the production process. The floor for the steam engines projects, and is flanked by large buttresses. A number of doors open to a corridor, enabling the workers to abandon the boiler room in case of violent gushes of high-temperature steam... As compared to the station of the Sector of Place Clichy, whose classicism witnesses the relationship with the Haussmannian urban fabric, the station designed for the CPAC is imposed as an autonomous architecture which clearly asserts its industrial function.

Friesé found his models in the industrial architecture of the period, turning his attention to the tradition of iron structure associated with walls in polychrome bricks, the examples of which were diffused by architectural magazines.[8] The load-bearing metal skeleton as a main feature, the use of new materials,

4

5. P. Friesé, electric sub-station "Voltaire" on Avenue Parmentier, Paris, 1908. Period photo. IFA.

6. P. Friesé, electric sub-station "Auteuil", Paris, 1912. Period photo. IFA.

the inventive arrangements were not sufficient to dispel the reminiscences of the "great palace" evoked by this station. The U-shaped plan, the wall structure with its combination of red and yellow brick – a variation of the so-called "Louis XIII" cladding – the vertical order of the openings of doors and windows, the continuous bosses of the foundation, the angular arrangement of the chains, the pavilion roof with tiles undoubtedly reveal a firm intention to adapt to the context.

Since power stations built after 1900 were distinguished from the "metallurgical aesthetes," due to the greater attention to design aspects, Friesé's nationalistic rhetorics ventured on an unexpected itinerary with the construction of about twenty electric transformation sub-stations. By dividing the Parisian distribution network into small square sectors, the sub-stations served as relays from the main stations, in order to attenuate the formidable AC power, deviating it in a more docile DC current. The model of this new type of buildings was established in 1903, with the construction of the sub-station of Opéra in Rue Caumartin:[9] a vast engine room on the ground floor, and a hall housing accumulators on the first floor. This series of sub-stations is characterized by its constructive scheme: a steel structure supporting the translucent façade consisting of an iron skeleton and glass tiles. The tripartite elevation, with the frame of the windows of the attic, finds an unexpected counterpart in the reinforced-concrete façade of the garage in Rue Ponthieu (1905) by Auguste Perret (1874-1954). Im both cases, the design of the façades reveals the architect's constructive adherence to the model of the Gothic cathedral. The longitudinal section of the nave, with its superimpositions of big arches and three-mullion windows, has thus been transposed to the façade and, while Friesé used the round arch, Perret geometrized the structure and integrated the principle of the rose-window in the main span. All the aesthetic interpretations derived from Viollet-le-Duc have therefore been summarized in this abstract reinterpretation of Medieval forms adapted to modern requirements.

"The Sanctuary of Energy"

We just have to recall Friesé's most famous work, which no longer exists, the one the lyrism of a critic had elevated to the level of "sanctuary of energy." The enthusiasm of his contemporaries was such that the Société des Artistes Français – a decidedly conservative organization – awarded him the gold medal at

5

6

the Salon of 1904, at the same time admitting industrial architecture to the elite of the major architectural disciplines.

A solitary creator, Paul Friesé certainly did not pay less attention to foreign electric architecture, especially German. Perfectly bilingual thanks to his origins, he visited Germany twice or three times a year, on business trips whose results he did not hesitate to make available to the French information services in his capacity of reserve officer. As of 1898 he summed up his first realizations and developed a completely nėw spatial organization for the construction of the enormous electric traction station of the Paris Subway, at the Quai de la Rapée at Bercy (1899-1904).[10] What he has drawn in his album under the initials UB ("Usine de Bercy" but also Untergrundbahn, subway in German) is the conclusion of a reflection on the "linear" station which has borrowed the division of the ancient Christian church basilica. Two significant buildings, the power station of La Triphase society in Asnières (1898-1901) and the Schneider electric engine factory at Champagne-sur-Seine (1901), illustrate the architectural declinations of this basilica theme. Lateral, lower aisles depart from the two sides of a tall main nave. Five naves of identical length are divided into regular spans which correspond to as many units of energy production: the coal deposits, located below, feed the boilers of the aisles which, in turn, supply energy to the electric turbines of the great central nave. While the construction's analogy with religious architecture was greatly emphasized at the time, the architectural casing of this "station-cathedral" gave rise to questions. Friesé has intentionally concealed the metal skeleton with a masonry structure in limestone brick imported from Germany. The same construction decisions can be found in the Bastille sub-station (1911), the most important of its category.[11] Friesé was not a man who limited himself to just one formula. The "modernity" of an architecture cannot be reduced to the structural assertion of any steel skeleton, thus prefiguring the modern station designs built in the period between the two wars. The reference to brick casings consecrates a research for architectural expressiveness evident in the geometric rendition of the volumes, in the play of solids and voids, in the oversized openings, in the soft moulding of the bricks with rounded corners, etc. This is not surprising, since the rigorism worthy of Viollet-le-Duc is superimposed by a romantic style, a Medieval revival that emphasizes the image of the station in a self-glorifying way. Friesé has here proposed a personal reinterpretation of the "Rundbogenstil," German neo-romanticism. The Bercy station has probably been inspired by the German churches that witness a precocious interest in the ancient Christian models – as the Friedenskirche of Potsdam (Ludwig Persius, 1843-8) or the Bonifacius Basilica of Munich (Georg Friedrich Ziebland, 1835-50). As Hubert Damisch has observed concerning the problem of historicism in Germany, "Rundbogenstil, the style characterized by round arches [...] is inspired, without distinction, by ancient Chris-

7, 8. P. Friesé, electric sub-station "Bastille", Paris, 1911. Fronts at the corner of Rue de l'Arsenal and on Boulevard Bourdon. Period photo. IFA.

7

8

tian, Byzantine, Romanesque, Italian Gothic art, and above all that of 15th-century Tuscany."[12] While Friesé is presented as an advocate of the integration of the Romanesque vocabulary in the Paleo-Christian syntax, his architectural language directly evokes Oriental models. The main façade of the Bercy station is clearly inspired by the *madrasa* adjoining the Persian mosque, with its monumental arch and the two smokestacks placed on the façade like as many *eyvân* and minarets.[13] The sub-station of the Bastille, on its part, evokes Islam funerary architecture, revealing a certain affinity with the mausoleum of the Samanides in Bukhara (central Asia). We find the same square plan, the central opening repeated on each façade, the rounded buttresses at the corners surmounted by ovoidal cupolas, the upper arches. Friesé did not attempt to reproduce its models with precision, preferring to make a free combination of purified elements. Having freed himself from the requirement of archaeological coherence, he has here been able to use historical references as starting points for a constructive analysis.

The oriental component of Paul Friesé's studied eclecticism should not come as a surprise in a period when, in France, Byzantine architecture was represented by the "very Christian" style of the Spanish church, for which Islam architecture had provided, for example, the model for the crematorium of the Père Lachaise cemetery, a true "Republican" mosque (J.C. Formigé, 1887-9) and for which the leader of the rationalist school, Eugène Viollet-le-Duc himself, wrote the introduction to the collection of writings on Turk and Arabic art.[14] Friesé's journey to the East was thus limited to an interaction of a bookish kind, namely the reading of the numerous collections published in the 19th century, following the archaeologic and scientific missions. And while Viollet-le-Duc dedicated a part of his 15th *Entretien* to Islamic architecture, the Paris architect mainly drew inspiration from the volume *Monuments modernes de la Perse* (1867) by famous architect, explorer and archaeologist Pascal-Xavier Coste (1787-1879).[15]

The Bercy station and the Bastille sub-station were contemporaries of the rational school which centered on Anatole de Baudot, the prophet of reinforced concrete, who exhibited projects at the Salon de la Société Nationale des Beaux-Arts characterized by an excessive predilection for the dome, the terrace roof, the lowered arch and polychrome decoration.[16] But it is with the project for the theater of the Champs-Elysées, by Belgian Van de Velde, that the Bastille sub-station finds an unexpected term of comparison, a juxtaposition which is even more surprising if one considers that the two projects are contemporary. Henry van de Velde (1863-1957), after a journey to the East in 1903, worked in Paris on what was to become the theater in Avenue Montaigne in 1911, before he was marginalized by Auguste Perret. His last sketches multiply the quotations of Oriental architecture, such as the dome placed on the terrace roof, and the entrance

9. P. Friesé, electric sub-station "Bastille". Detail of the main front. From: H. Fiblec, *Paul Friesé 1851-1917* (Paris: IFA/Norma, 1991).

9

marquee.[17] The buildings designed by Friesé and Van de Velde share several characteristics: the arched frames or doors and windows in the attic and the corner solution, with a semicircular tower covered by a dome which dovetails in the projecting walls. His encounter with the East made Van de Velde definitively abandon Jugendstil in favor of "pure form," while Friesé saw it as an opportunity to invent a new style that could correspond with a problem of "outdatedness": that of the "metallurgical" or Medieval-inspired style associated with the formal absolutism of industrial architecture. This architectural syncretism is also an acknowledged expression of a Sansimonian Freemason who unites East and West in a chimerical pursuit of the universal.

The symbolic charge of Friesé's electric architecture is not casual, if one considers that the constrast between light and darkness is a recurrent theme in Freemason rites. Electricity violates the millenarian order imposed by the solar rhythm. "And by night," writes Emile Zola, "in the black sky, [electricity] will light another sun which will extinguish the stars. And it will suppress winter, and give rise to eternal summer [...]."[18] The age of darkness will be followed by the age of light. If the Electric Fairy is the religion of the 20th century, the power station becomes its sanctuary.

1. E. Zola, *Les quatre évangiles. Travail*, 1901.
2. On the complete work of Friesé, see H. Fiblec, *Paul Friesé, architectures de l'age industriel* (Paris: IFA, Norma, 1991).
3. The six licensee companies were: Compagnie Continentale Edison, Société Anonyme d'Eclairage et de Force, Compagnie Parisienne de l'Air Comprimé, Compagnie Electrique du Secteur des Champs-Elysées, Compagnie Electrique du Secteur de la Rive Gauche, Société du Secteur Electrique de la Place Clichy.
4. Power station of the Société du Secteur Electrique de la Place Clichy, 51-53 Rue des Dames, Paris, 17th arrondissement. Architects Paul Friesé and Jules Denfer. The façade of the office building is recorded in the Inventaire Supplémentaire des Monuments Historiques.
5. The architectural rendition of the power station of the Place Clichy Sector has an English counterpart in the Power Station of the Pall Mall Electric Light Company, Davies Street, London (Charles-Stanley Peach, 1889-91).
6. Power station of CPAC, 132-134 Quai de Jemmapes, Paris, 10th arrondissement. The building, property of the Société Papetière Clairefontaine, is recorded in its entirety in the Inventaire Supplémentaire des Monuments Historiques.
7. This daring superimposition, made delicate by the installation of steam engines above the electric generators, which are not protected from possible water leakage, has a precedent in the power station of Deptford in England (F.Z. de Ferranti, engineer, 1887-91). The two are nevertheless distinguished by the fact that in the case of the latter it is merely a matter of the superimposition of boilers.
8. Mill of the Menier chocolate factory in Noisiel in Seine-et-Marne (Jules Saulnier, 1871-4); barracks for the Garde Républicaine in Rue de Schomberg in Paris (Jules Bouvard, 1878-83); a plant for compressed air of the CPAC (Joseph Leclaire, 1890-1).
9. Electric sub-station "Opéra" by the Compagnie du Métropolitain de Paris (1903-4), 41 Rue Caumartin, Paris, 9th arrondissement, still in use. The façade on the street is recorded in the Inventaire Supplémentaire des Monuments Historiques.
10. Electric traction station of the Compagnie du Métropolitain de Paris, 46-52 Quai de la Rapée, and 177-179 Rue de Bercy, Paris, 12th arrondissement, now demolished.
11. "Bastille" sub-power station of the Compagnie du Métropolitain de Paris, 31 boulevard Bourdon-18 Rue de l'Arsenal-6 Rue de la Cerisaie, Paris, 4th arrondissement. The building, which is still in use, is recorded in its entirety in the Inventaire Supplémentaire des Monuments Historiques.
12. H. Damisch, "Historicisme," in *Encyclopaedia Universalis* (Washington, 1980), vol. VIII, p. 449.
13. Friesé seems to have been directly inspired by the mosque of Mesdjid-i-Chah or Masjed-é Châh (mosque of the Shah), the great mosque at Isfahan erected in the early 18th century by a sovereign of the Safawid dynasty. It was published by Viollet-le-Duc in his 15th *Entretien*. The Karlkirche of Wien (1716-38) by J.B. Fischer von Erlach, pictured between two columns (evoking the Trajan column), is another plausible source of inspiration.
14. J. Bourgoin, *Les arts arabes*, preface by E. Viollet-le-Duc (Paris: Morel, 1873); L. Parvillée, *Architecture et décoration turques au XVe siècle*, preface by E. Viollet-le-Duc (Paris: Morel, 1874).
15. Also Viollet-le-Duc refers the readers of his *Entretiens* to the work of P.-X. Coste, *Monuments modernes de la Perse mesurés et décrits par...* (Paris: Morel, 1867). The illustration of the mosque of Isfahan contained in *Entretiens* has been taken from the collection of Coste.
16. M.-J. Dumont, "The Philosophers' Stone: Anatole de Baudot and the French Rationalists," in *Rassegna* (*Reinforced Concrete: Ideologies and Forms from Hennebique to Hilberseimer*), No. 49, March 1992, pp. 36-43.
17. M. Culot, "De la scène tripartite au théâtre total," in *Henry Van de Velde, projets du Théâtre, 1904-1914* (London: AAM-Architectural Association, 1974), p. 20.
18. E. Zola, *Les quatre évangiles* cit.

Ornella Selvafolta

The Power Station, the Client, the Architect

"Wherever they go Italians encounter lines that run along or cross the countryside, the roads and the rivers. In the valleys they find pipes climbing the mountainsides, they admire new buildings without the characteristic smokestack, they notice smokestacks that lack their plume of smoke..."[1] These were the indisputable signs of the progress which the "Associazione fra Esercenti Imprese Elettriche in Italia" referred to in the preface of a luxurious publication on the country's most important power stations: a landscape dotted by unusual presences and innervated by an uninterrupted net of transmissions, where the wires which "meet in all directions" create "singularly fascinating" sceneries of "unusual movement."[2] Apart from the aesthetic considerations suggested by those interested in electrifying the territory, obviously inclined to discover picturesque values in the often chaotic routes of the power lines, these considerations introduced an element in the national scenario which was becoming the most characteristic and contemporary one as well as the most significant premonition of future developments.[3]

The web of wires that animated the natural and built scenario, supported by towers, stretched on iron footbridges crossing roads, canals and railways and assembled near the cities actually originated from remote locations where the *vis mechanica* was drawn from the sources to be transformed into electrical power which could be channeled into the distribution system. What had, in the late 19th century, triggered an extraordinary change in the production and consumption of energy was not, in fact, electricity as such, but the fact that it was now possible to transport it over ever greater distances, in order to connect, and approach, conceptually speaking, remote and different locations and make the territory of design virtually inexhaustible.

The energetic map of the country had indeed changed radically since, in 1892 and 1898, the power stations of Tivoli and Paderno had demonstrated that the waterfalls of the Aniene and the waters of the Adda, converted into electrical "inexhaustible shudders," could travel twenty or thirty kilometers, reaching Rome and Milan where they made the streetcars run and powered the machinery of the industries built along their routes.[4] Most importantly, Tivoli, Paderno and their successors were proof of a kind of national record; they witnessed that

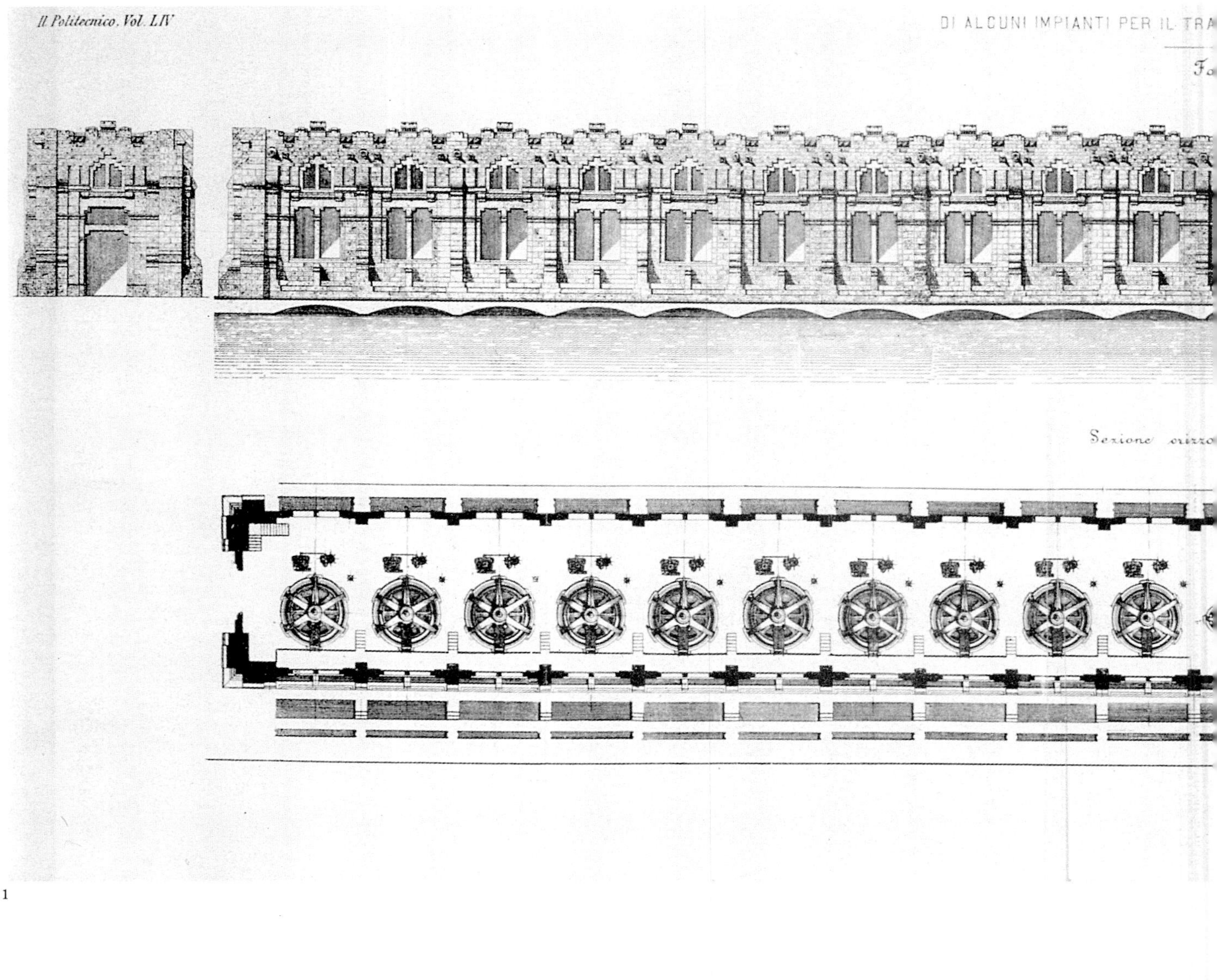

1

2

1. G. Moretti, Taccani power station at Trezzo d'Adda, 1905-1906. Front and plan.
Form: *Il Politecnico, giornale dell'Ingegnere Architetto Civile ed Industriale*, vol. LIV, Milan 1906.
ENEL Historical Archives, Milan.

2. G. Moretti, Vigevano hydroelectric plant for the Conti Electric Company, 1905-6.

the country was a forerunner in the European scenario in the field of transporting electricity over long distances: the result of a *savoir faire* characterized by a combination of the ancient hydraulics tradition and a rooted construction know-how, consolidated in the progressive creation of an "electrotechnical family" which also kept abreast with entrepreneur developments[5] and could draw from unheard-of natural resources. The *houille blanche*, as Emirico Vismara wrote on the occasion of the "First International Conference on Motive Power" (London 1924), in fact abounded in the country thanks to its numerous ridges and irregular rivers, and it was therefore possible to partially fill the energetic gap which had opposed it to the northern countries (rich in "black coal") and to launch a "fruitful hydroelectric spirit" that made up for past defeats and a long and compulsory subjection.[6]

At the end of the century little more than 500 km of transmission lines had been installed; at the end of World War I the figure had grown to 25,000 km, and a "fascinating" web of aerial cables had been stretched all over the peninsula, creating connections, deviations and links, to the point of "forming a vast and organic general plan" where "rivers were united, different hydrologic regimes were regulated, excesses and deficiencies of water in the various regions were compensated."[7] This system of connections and junctions appears even more interesting as it is associated with the destinies of many electricity companies, which were first founded in an exponential way in the period between the late 19th century and the first decade of the 20th century, and then reorganized by mergers and spin-offs on the basis of a complex but efficient network of business joint ventures and mutual economic interests.[8]

All this represented the starting point also for the particular phenomenon of the architecture of power stations, which took form and developed precisely in coincidence with the progress of the transmission of energy over long distances. In fact, this gave rise to a leap in scale and quality of the category and physiognomy of the stations, replacing the small, numerous installations "scattered along the watercourses, each using its own section" of mechanical energy with large stations that received, transformed and distributed, located in the "most profitable" natural locations[9] and whose *facies*, so to say, in turn, represented a national pride,

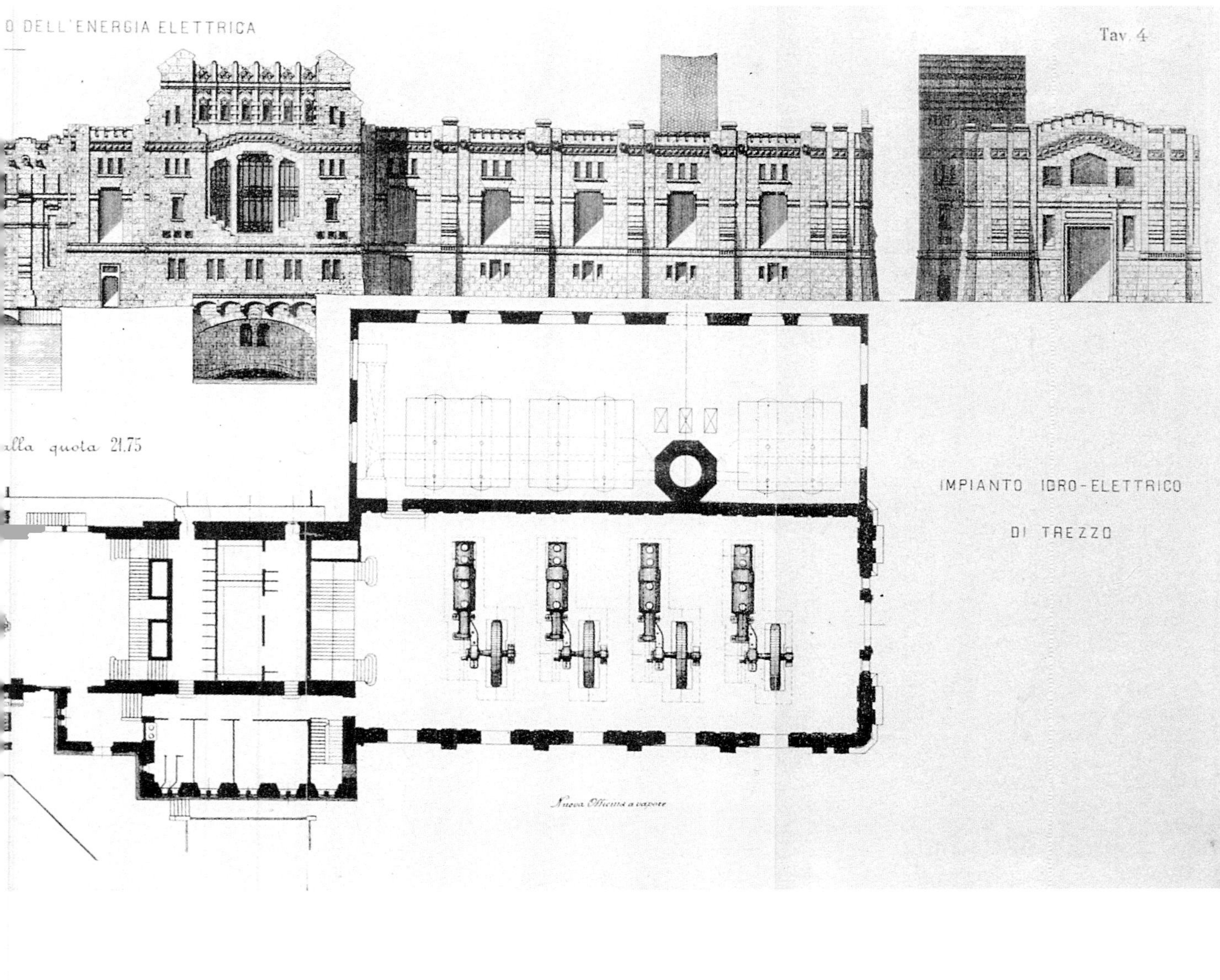

were in fact emphasized by Vismara at the aforementioned conference in London.

Many of the Italian stations stood out, he wrote, thanks to a particular "aesthetic line" and an unusual research of quality which made them lasting "monuments" of the "new Italy of the centuries to come."[10]

The power stations of the first decades of the century, new buildings in the sense that they were technical and functional novelties, however interpreted the "electrical revolution" with the expressive means characteristic of the culture of the period, uniting an interest in antiquity and modern ambitions in which the qualitative treatment of the image, or rather the conscious attention for the building as such, and its significance for purposes of advertising and promotion, expressed the sense of the innovation. In other words, it is pointless to seek, in the hydroelectric stations of those years, the linguistic revolutions and the avant-garde contents which the futurists, "hymnists of energy," militated for in the same period; it is useless to look for lineages with Sant'Elia's splendid drawings, characterized by their industrial conciseness and integration between the oblique and dynamic lines of the building and the rhythm of a technological scenario crossed by the forced pipes and buttresses of the dams; the stations would rather tend, in the best cases, to adapt the echos of the architectural traditions to new functions and to express the naturalness of the locations and their landscape values, rather than the technological aspects.

One of the best cases certainly is the hydroelectric station which was to become an obligatory stop in the itinerary of electric architecture: the station of Trezzo d'Adda of 1905-6 whose machinery, planned by engineers Adolfo Covi and Alessandro Taccani, was given an external appearance by the skilled hand of architect Gaetano Moretti. Considered by recent critics as one of the best works in a professional career that wavered between late eclecticism and modern inspirations,[11] it was appreciated not only for its electrotechnical part, but also for its architecture *tout-court*, the example of an Italian approach to modernity, where it was possible to integrate novelty with tradition and find a balance between the uncertain limits of memory and design.

This is the only feasible explanation of the attention paid to a type of building which generally speaking did not belong to the categories reviewed by the magazines and authors specializing in architecture: in 1909 Ugo Monneret de Villard (who, incidentally, designed, in turn, the architectural side of hydroelectric

3

3. P. Portaluppi, Verampio hydroelectric plant for the Imprese Elettriche Conti, 1913. Power station. Portaluppi archive, Milan.

stations)[12] included it in *Opere di Architettura Moderna*, written on the eve of the Architecture Exposition of Vienna, while the magazine *L'Edilizia Moderna*, in the same year, devoted an article to it.[13] The authoritativeness of architect Moretti, the favorite pupil of Boito and his successor as professor of architecture at the Milan Politecnico and at the Brera Academy, obviously counted, but the most decisive factor was that the hydroelectric station, as it was conceived, assumed a representativity considered intrinsic of the best architecture, regardless of the separation, which was on the contrary considered determining by later critics, between science and art, between technical and aesthetic solutions, between engineer and architect.[14]

We thus witness the singular circumstance that, while the quoted sources discussed the station mainly in terms of formal project, the much more numerous and detailed writings describing it in the specialized technical magazines on the contrary abound in reports of another kind; here the water capacity, the spillway and the tailrace, the walls of the dam, the hydraulic turbines with vertical axes, the alternators, the dynamo, the multitubular boilers, the control panels, the ring circuit and so on are discussed exhaustively, often neglecting or even omitting the name of the architect.[15]

This is, however, not the only such case: the station of Vigevano, built for the Società Conti (1906), the thermoelectric station of Novara and the hydroelectric one of Piedimulera (both started in 1907 for the Società delle Forze Motrici dell'Anza), also designed by Moretti, were published without mentioning their architectural author, while also in the 1920s the highly specialized magazines of the sector, which described the stations more and more in detail, showed the same lack of interest, proof of a more than accepted separation of roles where the artistic one was frankly considered marginal for purposes of efficiency. While the technicians showed little sensitivity towards the "aesthetic line" which was to be considered the prerogative of Italian stations, the responsibility for the choice of images lied with others, and in the case in point, with the client: the chairmen, managing directors or general managers of the numerous electricity companies which, in the early years of the century, competed for the exploitation of hydraulic power in a very aggressive environment. Often neglected in the history of electrical architecture, the clients were on the contrary, at least as far as important figures like Moretti and Portaluppi are concerned, often

4, 5. P. Portaluppi, Verampio power station. View of the powerhouse, the garden and the manager house. Bird's eye view of the power plant. ENEL Historical Archives, Milan.

involved in a relationship with the architect that went beyond the mere assignment of the artistic rendition of the station.

This is the case of Benigno Crespi, a well-known cotton manufacturer and chairman of the Società Anonima delle Forze Idrauliche of Trezzo sull'Adda founded in 1904 to exploit the power of the river, to power his industries and distribute energy in the Milanese district. In his capacity as Director of the Regional Office of Monuments, Moretti had already somewhat conflictive relations with Crespi, concerning the issue of the Castello Visconteo of Trezzo, which the industrialist had bought in 1892, with the intention of making alterations that failed to respect its historical significance. However, when the architect won the competition for the cemetery of the workers' village of Crespi d'Adda,[16] the old dispute was forgotten, and a few years later the assignment for the architectural part of the Trezzo station set the seal on a renewed relationship of collaboration.

The objective of this project was to create a suitable appearance for the "energetic heart" of Crespi's feud: a building installed on the banks of a river in a particularly picturesque stretch, dominated by the very castle which had previously given rise to dispute; but it was also a question of doing justice to a masterpiece of engineering where "the turbine hall was integrated with the intake building" and the water arrived directly to the engine room without resorting to the complex series of hydraulic buildings which marred the landscape with artificial loading tanks, forced pipes, diverting canals. In other words the building represented an "immense lung" which drew water under pressure, rather than making use of the force of gravity, channeling it back to the river on the opposite side of the promontory through two underground tunnels.

This solution, interesting from a technical point of view, was also charged with expressive potential, since the romantic landscape of the Adda, which made a double curve between the banks dominated by the rocks and by the castle ruins was, it is true, altered, but by a sympathetic eye and a light hand. As *L'Edilizia Moderna* explains, Crespi himself was not foreign to this approach. An advocate of a modern design that blended harmoniously with nature and history, he in fact explicitly asked the architect to "harmonize with the surrounding nature and to avoid creating conflicts with the ruins of the ancient castle above."[17]

Correctly arranged according to the stages of production (turbine room, control center, thermal workshop), but also rooted in the pro-

4

5

6. P. Portaluppi, Crego power station for the Imprese Elettriche Conti, 1917. Banca Immagini ENEL.

found historical culture of its author and respectful of the surroundings, the Trezzo station thus presents itself in its "apparel" of rustic lineage, corrugated by projections and niches which leave room for the openings and cling to the arrangement of the machines, dialoguing with the irregularities of the rocks, projecting indentations which allude to the battlements of the castle above, immediately guiding the eye towards it, showing Medievalistic motives that merge with more modern designs, connecting the artificial with the natural, harmonizing a world of "opposites" where, as Luca Beltrami writes, the "memory of the past" could co-exist with the "intuition of the future."[18] "Memory and intuition," a kind of double register which we know were a favorite of Italian architecture of the period,[19] and, as a reflection, of its "electric declination."

Ettore Conti, as a young engineer graduated from the electrotechnic department of the Milan Politecnico, was very aware of the "electric prediction" which was, in his opinion, "in the air"; in 1901 he thus embarked on a very brilliant career, founding the homonymic Società per Imprese Elettriche (which in 1911 already participated in at least twenty companies operating in the field of distributing energy) and working (also in his capacity as a Member of Parliament) in favor of an integrated policy of energetic exploitation of the country.[20] Ettore Conti was a tireless animator of the latter, a fact which became evident in 1948 when his notebook, full of initiatives and commitments which also reveal interesting information on power stations, was published. It reflects a sincere interest in the architectural design of the buildings, evident already in the first constructions: the station of Zogno, on the Brembo river (1903) and the station of Vigevano, also entrusted to Moretti (1906), both characterized by "neo-15th century grace," with projecting or round arches, profiles or roses in terracotta, small balconies, graffiti.[21] "I am pleased to have given the stations also an artistic appearance," Conti wrote in this regard,[22] also revealing a trace of his predilection for the Lombardesque, permeated by references to late-Gothic and to Bramante, which then belonged unconditionally to the formal universe indicated by Boito and Moretti at the Milanese polytechnic institutes and which, in the same years, received authoritative endorsement in the numerous restorations and the abounding studies on Lombard monuments.[23]

But Conti's interest in the central building was to grow as his development projects were realized, in particular those in the Val d'Ossola

6

in Piedmont, aimed at supplying energy to "the entire vast region including Novara, Vercelli, Alessandria, Novi, Tortona," rich in industries that were "very flourishing, but susceptible to further development."[24] "In the warm season I visited the valley for about ten days, only accompanied by Giannina [his wife] in order not to attract too much notice; I crossed the different level lines with the only help of my pocket barometer,"[25] he wrote down in his notebook, revealing that the "electrotechnical family of Italy" could benefit from intuition and discretion, an explorer spirit and a sharp eye for the nature and its potential riches. The favorable judgement of the "level lines" decreed by the "barometer," the ten days spent among the ridges "frosted by glaciers, with green pastures and brown spots of nude rocks [...] in the thick of the ancient coniferous forests"[26] were, as we know, to lead not only to the complete exploitation of the catchment basin of the Toce and Devero rivers, but also to a fertile season of hydroelectric stations, whose architectural design was entrusted to Piero Portaluppi, a pupil of Moretti and successor to the latter's architecture professorship.[27]

Endowed with an "agile talent" that enabled him, to a greater degree than others, to harmonize "the prestige of the force [...] with the smile of beauty,"[28] he and Conti began a stable collaboration which, while certainly facilitated by family relations,[29] was nonetheless significant and propitious, due to both the indisputable intellectual caliber of the two, and the abundance of opportunities which were to appear both with regard to stations and other buildings, such as the Casa degli Atellani in corso Magenta in Milan, the Conti-Sayno shrine at the Milan Cimitero Monumentale, the design of suburban villas and the restorations, financed by Conti himself, in Santa Maria delle Grazie church, also in Milan.

From 1913 until 1926 (when the Società Conti was incorporated with Edison) Portaluppi designed six hydroelectric power stations in Val d'Ossola, leaving evidence in the numerous drawings and photographs published in the two volumes of *Aedilitia*:[30] a true constellation of plants, or rather, as was emphasized, a "necklace of gems" set in that nature whose beauty the electrotechnicians were, in the final analysis, able to evaluate well, while turning it to their advantage. The "best gem" of this series was, according to Conti, the first: the station of Verampio, designed by Portaluppi three years after graduating (in 1910) and certainly the one which was most influenced by the taste of the client. It is, in this regard, worth while to quote some comments from the "notebook": "Long discussions with architect Portaluppi on the architecture to adopt for the Verampio Station [...]; where the Toce and Devero rivers meet, it will represent, to some extent, the entirety of the complex system of exploitation of the catchment basin, and I want it to assert, also externally, the spirit which has animated me in this venture. We naturally must not be tied to any style belonging to the past; nor would I like it to be one of those buildings which are so fashionable today, whose simple lines and precious materials are badly suited to the characteristic dignity of the valley; the materials must absolutely be local, and not only for economical reasons; therefore, granite and gneiss stone; no terracotta, no marble. As a general line I would like to base the design on that of the church of St. Moritz, which I made a sketch of last winter for this specific purpose."[31]

Built at the foot of mount Cistella, overlooking the river bank, the station indeed assumed some of the characteristics of its model in the large portals and sloping roofs, but it became, above all, a towered castle, or rather an impressive fortified villa facing a symmetric garden, situated in the center of a "village" (the buildings housing the transformer, the workshop, the dwellings of the director and shift foremen), modulated between the Neo-Romantic compactness of the masses and the materials, the delicacy of the early Renaissance elements and the modernity of contemporary details; a faithful rendition of the client's wish, but also a clear derivation of the art of composing with a mastery of style which Portaluppi, having studied under Boito and Moretti, had exercised when attending the Politecnico.

All this represented a strong symbolic significance to Conti: this was in fact the "heart" of the future stations of the Ossola area, a barycentric location where the rivers converge, where cables gather and branch out, a sample of the great energetic plan which the entrepreneur intended to realize in the area, as tangible demonstration of a technical project and business proposal in which the exact conceptual arrangement could achieve the beauty of a "work of art."[32] Enlightened industrialist, sincerely convinced of the great dignity and responsibility of his role, Conti saw himself as a kind of new Renaissance "prince" burdened by economical obligations and cultural duties and entitled to the functions of power and those of distinction and style. In fact, when Portaluppi in 1919 proposed to buy and "restore" the ancient Casa degli Atellani he immediately agreed, entrusting him with the task of arranging it as a "comfortable and artistic residence," and following the works with "great love": the proposals of inscriptions and paper decorations, the surfacing of old frescoes and the invention of new ones, the planning of sundials and zodiacal signs, graffiti, mosaics, wrought iron; experiencing "enjoyment and particular cause of satisfaction" in planning a Renaissance substantiated by history, but also cloaked by the amiable veil of the tale.[33]

7

7. P. Portaluppi, Molare power station for the Officine Elettriche Genovesi, 1920.
ENEL Historical Archives, Milan.

8

8. P. Portaluppi, Valdo power station for the Imprese Elettriche Conti, 1922.
Banca Immagini ENEL.

This kind of amiable veil was also to cloak Portaluppi's subsequent stations: a light veil that enabled the "electrical fable" to abandon the somewhat monumental rhetoric that could still be observed at Verampio, in order to improve in terms of vivacity and imagination, almost as if the client, by now satisfied with his "private Renaissance," had gradually come to understand the more creative vein of the architect. In fact, we no longer find any traces of "discussions" and disputes in the notebook, but only the evidence of the entrepreneur's satisfaction in the realized works, his delight in the thrill elicited both thanks to the technical solutions and the architectural lines, the notes on visits by Italian and foreign authorities, by king Vittorio Emanuele III.

The "necklace" was also enriched by "jewels" such as the stations of Crego, Valdo, Sottofrua, Crevola-Toce, Cadarese, all of them started between 1922 and 1928 and obviously associated with an impressive series of land works: the transformation of alpine lakes into basins, the building of pipes, dams and barrages, of tunnels for filling and emptying, spillways and water intakes, stabilizing basins and impact-absorbing constructions, buildings for the registration of capacity. There is an abundance of technical articles discussing these aspects, featuring formulas and figures which in some cases assume mythic dimensions and are subject to bombastic emphasis, evoking the futuristic praises of technological progress. "Natural power," "dreamed energy which follows the itinerary of human genius," "benign fairy who brings light, work, heat," "one of the grandest and most poetic phenomenon of the eternal vicissitude of cosmic energies": these phrases ring with the lyricism the engineers revelled in,[34] echoes of Marinetti's "electrical war." Conversely, and this is why they are interesting, the buildings in which the "powers of the cosmos" met and their magic alchemic transformation took place, succeeded in transmitting the dignity of the phenomenon without the solemn attire of rethoric indulgence.

The most striking characteristic in Portaluppi's stations – which becomes more and more evident – is his own personal poetics which, while complying with the client's wish to respect the "characteristics of the valley," nevertheless reflects also an inclination to variety and movement which makes them harmonize with the "flow" of the Alpine panorama. "Granite and gneiss stone," rough-hewn stones, exposed wood structures, broken profiles, indentations, set-offs, an alternation of rough and smooth surfaces, thus commented the animation of the picturesque mountain landscape in the Crego station,[35] which was built in the midst of a thousand difficulties during World War II. It faced the Toce river under the lee of a steep rock in "granite and crystalline shale" which had almost certainly "inspired" the angles and slopes, the increasing density of scratches in the irregular *opus*, the pointed edges, the steps cut into the gable ends. While it still, to some extent, was inspired by the idea of the castle (and in the same period Portaluppi built a fantastic "little castle" in Valtellina for the Roasco station of the Azienda Elettrica Municipale of Milan), Crego already showed a predilection for linguistic contamination,

9. P. Portaluppi, Crevola power station for the Imprese Elettriche Conti, 1925. The engine room.
ENEL Historical Archives, Milan.

10. P. Portaluppi, Valdo power station. The engine room.
ENEL Historical Archives, Milan.

11. P. Portaluppi, Cadarese power station for the Edison Company, 1925-8. The engine room.
Banca Immagini ENEL.

12. P. Portaluppi, Crego power station, 1917. The engine room.
Banca Immagini ENEL.

where the echoes of tradition merged with secessionist, Futurist and Art Déco influences, emancipating from the "tyranny" of the exact copy or the didactic transposition of styles.

In the next station in the Ossola area, that of Valdo in the upper Formazza valley, which was started in 1922 and exploited mountain basins located at an altitude between 3000 and 2000 m (with a difference of no less than 700 m), all these aspects were to be condensed in a more refined quotation, turning into more concise stretches with stones and plasterwork, more clearly defined profiles, more abstract motives and fortunate decorative inventions,[36] such as the two rows of large "knobs" in faceted stone on the sides of the portals, the improbable bosses with graffiti interiors and the frieze in the engine room featuring pines that had been stylized, cut and planted in vases, a "treacherous" rendition in a location surrounded by "ancient coniferous woods."

These are symptoms of the ironic vein which has been emphasized on several occasions with regard to Portaluppi, of his amused and irreverent eye which he had already had the opportunity to exercise when illustrating the satyrical journal *Guerin Meschino*, and which in those years made him add bizarre caricatures here and there on the drawings for the very serious hydroelectric stations or which made him hint, not without subtlety, to the inevitable artificiality which technical installations in a natural context entailed. Also the Alpine architecture adopted for the dwellings around the station appear, deep down, as intentionally artificial, elevating the repertory of the picturesque to an art and to a fairy-tale dimension, for the houses destined to technicians able to interpret the hum of the turbines, to read

9

10

11

12

13. P. Portaluppi, Molare power station, 1920. The engine room. Portaluppi archive, Milan.

panels and indicator lights, to open and close circuits, to press buttons and free energy along the paths "followed by human genius."

It may be one of them who, in the drawing of a detail on the portal of the Crevola-Foce station, strikes by lightning an improvident interlocutor;[37] an irascible technician on a threshold "embellished" by two large ceramic insulators, conceived for Portaluppi's most interesting and linguistically most innovative electric building.[38] In fact, in 1926, Gaetano Minnucci, when publishing an article on "the architecture and aesthetics of industrial buildings," chose a photograph of the station for the first page, proceeding with descriptions of buildings by Fahrenkamp, Gropius and Dudok, and denouncing the mistake of the duality between technical and artistic part which had, as a rule, characterized this kind of realization, and decreeing the end of the castle-inspired and embattled industrial plants.[39]

Leaving behind any mimetic flattery with respect to its surroundings, Crevola (not oblivious of the example of Trezzo) was in fact organized in three distinct volumes, namely the generator workshop, the water-cooling tower and the transformation sub-station, with rectangular openings, elongated heightwise in order to make room for the inlets to the engine room, bridging the outlet canal, extending horizontal loopholes for the cables, calibrating the decorations according to a more marked abstraction and geometry. However, strictly speaking, it does not seem to belong to the cases of *exempla virtutis* formulated by Minnucci on the basis of a new "architectural art that is [...] mathematical, strong, rational, rigidly logical,"[40] as it also gives room for fantasy, distortion, unexpected combinations. Among the slopes of the Ossola the water-cooling tower in fact assumes oriental, light traits evoking the pagoda; the rusticwork becomes lozenge patterns, the windows are set as flaming rhombs in the plasterwork, the roof projects on the corners, the "Serlianas" increase in width to follow the grid and patterns of the rusticwork, the gables reach upwards with pointed extremities, the marble, previously excluded, faces the portal and its syncopated pediment. Everything suggests that this is where lighting is produced, but the serious chords accompanying the "celebration of the bolt" are absent, and the "agile genius" which Conti had perceived in the architect's earliest works, remains. The technician who, on the main entrance of Crevola, strikes someone by a lightning evokes Gadda's engineer, who would only part with his slide rule when he's taking a bath in the sea,[41] and likewise Por-

13

14. P. Portaluppi, Cadarese power station.
Banca Immagini ENEL.

taluppi's stations may appear as an architectural transposition of the writer's prose with his irresistible vocation for the deformation and mimetic amusement, for contamination and for the catalogue and the inventory. Both look, moreover, with disenchanted and somewhat mocking eyes at the tools of their own work: Portaluppi, the architect, was thoroughly familiar with history, styles and consonances, thus being able to alter them without lowering the quality, even on the verge of parody, while Gadda, the writer, mastered the word equally well, being able to soar over his immense territory, hosting special languages, cultured idioms and dialects, while Gadda, the engineer, also knew technology, and could thus lift the veil of the myth and observe it with smiling irony.

14

1. Associazione fra Esercenti Imprese Elettriche in Italia (AEIE), *Notizie sui principali impianti elettrici d'Italia* (Rome, 1911), pp. 3-4. The AEIE was founded in 1899; in 1911 it began publishing the periodical *L'Industria Elettrica*; the members, who amounted to 60 in 1899, increased to 212 in 1911. It was chaired by engineer Carlo Esterle, and engineer Ettore Conti was its vice chairman.
2. G. Colombo, *Il trasporto dell'energia* (Rome, 1911), p. 19.
3. See O. Selvafolta, "L'immagine del paesaggio tecnologico nella Lombardia del primo Novecento," in C. Pirovano (ed.), *Lombardia. Il territorio, l'ambiente, il paesaggio*, vol. V, pp. 69-128, 285-8.
4. In general, on the history of Italian hydroelectric plants, see the four volumes of AA.VV., *Nel cinquantenario della Società Edison 1884-1934* (Milan, 1934). As for Paderno, in particular, see Società Generale Italiana Edison di Elettricità, *Note sui nuovi impianti 1895-1898* (place and date unknown, ca. 1899-1900).
5. Reference is made to the department of industrial engineering of the Milan Politecnico and, in particular, to the special school or specialization course in "elettrotecnica" made possible as of 1887 by a private donation (Istituzione Carlo Erba) which trained students of great competence. Cf. *Nel cinquantenario della Istituzione Elettrotecnica Carlo Erba, 1887-1937* (Milan, 1934).
6. E. Vismara, *Gli impianti idroelettrici in Italia e il loro contributo alla economia del paese*, First international conference on motive power, London, June-July 1924 (Milan, 1925).
7. *Ibid.*, p. 20.
8. An updated list of Italian electrical companies can be found on a regular basis in the magazine *L'Energia Elettrica*, published in Milan as of 1924 by ANIEL (Associazione Nazionale Industrie Elettriche). See, for example, the 150 companies registered as of 31 December 1925 in *Ibid.*, III, No. 2, February 1926, pp. 161-3.
9. The quotations are taken from G. Colombo, *Il trasporto dell'energia* cit., p. 12.
10. E. Vismara, *Gli impianti idroelettrici in Italia* cit., p. 10. See also E. Conti, "Lo stato attuale del problema idroelettrico in Italia," in *L'Energia Elettrica*, II, No. 2, February 1925, p. 149.
11. For an outline on Moretti, see L. Rinaldi, *Gaetano Moretti* (Milan, 1993).
12. He designed, e.g., the power station of Varzo on the Diveria river in Val Divendro: cf. "L'utilizzazione del bacino del torrente Diveria da parte della Soc. It. per Impr. Elettriche 'Dinamo' e gli altri Impianti della Società stessa," in *L'Energia Elettrica*, No. 3, March 1924, pp. 228-60.
13. Cf. U. Monneret de Villard, *Opere di architettura moderna* (Milan, 1909); "La centrale di Trezzo," in *L'Edilizia Moderna*, XVIII, 1909, pp. 81-2. On the importance of the power station as a "model of hydroelectric architecture," see also F. Secchi, "L'architettura delle centrali elettriche," in *L'Energia Elettrica*, IV, 1927, pp. 1331-8.
14. Moretti himself considered it his masterpiece, regardless of the fact that one had to attribute "the distributive definition and the rigid functional control to the engineer colleagues." L. Rinaldi, *Gaetano Moretti* cit., p. 179.
15. Among numerous technical articles, see A. Covi, G. Semenza, "L'impianto idroelettrico di Trezzo sull'Adda," in *Il Politecnico. Giornale dell'Ingegnere Architetto Civile Industriale*, vol. LIV, 1906, pp. 65-72, 211-2, 325-7, 357-9.
16. On this event see L. Rinaldi, *Gaetano Moretti* cit., pp. 156-7, 178-9.
17. "La centrale di Trezzo" cit., p. 81.
18. L. Beltrami, "La mostra della ditta Ceruti (architetto Gaetano Moretti) all'Esposizione d'Arte Decorativa Moderna di Torino," in *L'Edilizia Moderna*, XII, 1903, p. 31.
19. For a description of these trends see O. Selvafolta, "Il nuovo e la tradizione: aspetti dell'architettura del primo Novecento," in B. Zilocchi, M. Iotti (eds.), *Gli anni del Liberty a Parma: architettura dal 1900 al 1925* (Parma, 1993), pp. 11-7.
20. Cf. E. Conti, *Per una politica nazionale delle forze idroelettriche in Italia* (Milan, 1916); and also Idem, *Dal taccuino di un borghese* (Milan, 1948), 1st ed., here used in the 2nd ed. of 1971.
21. Cf. Eng. Villoresi, "Di alcuni impianti per il trasporto dell'energia elettrica. Impianto idroelettrica del Brembo, Centrale di Zogno," in *Il Politecnico. Giornale dell'Ingegnere Architetto Civile e Industriale*, LIII, 1905, pp. 321-32, 426-9.
22. E. Conti, *Dal taccuino...* cit., p. 34.
23. On the teaching of architecture at the Politecnico during the first years of the century, which correspond to the period when Piero Portaluppi studied there, see L. Rinaldi, "Gaetano Moretti e la formazione degli architetti tra eredità boitiana e rinnovamento," in *Il Politecnico di Milano nella storia italiana (1914-1963)* (Bari, 1988), vol. II, pp. 613-21; cf. also O. Selvafolta, "La scuola di architettura al Politecnico di Milano negli anni della formazione di Muzio," in *L'architettura di Giovanni Muzio* (Milan, 1994), pp. 25-35.
24. E. Conti, *Dal taccuino...* cit., p. 38.
25. *Ibid.*
26. The quotation has been taken from G. Ganassini, "L'impianto idroelettrico di Crevola d'Ossola sul fiume Toce," in *L'Energia Elettrica*, III, No. 8, August 1926, p. 636.
27. On Piero Portaluppi see the recent contribution by G. Bilancioni, *Aedilitia di Piero Portaluppi* (Milan, 1993).
28. E. Conti, *Dal taccuino...* cit., p. 62.
29. Piero Portaluppi's wife, Lia Baglia, was the

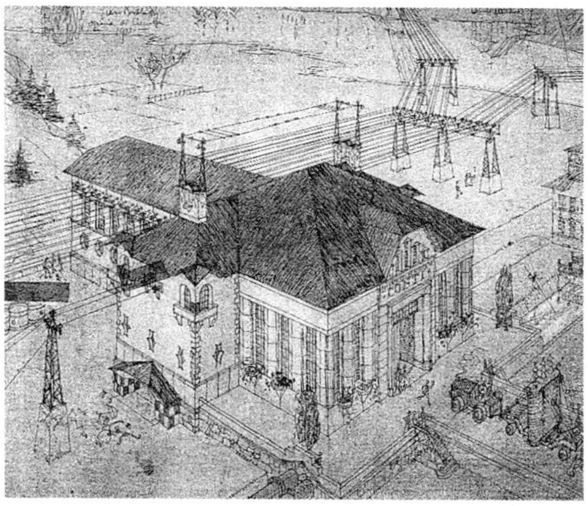
15

15, 16. P. Portaluppi, Crevola power station. Preliminary sketch, perspective view, 1923. Portaluppi archive, Milan.

17. P. Portaluppi, Crevola power station, 1925. Banca Immagini ENEL.

niece (the sister's daughter) of Ettore Conti and was adopted by him in 1939 together with the other nephew Piero Gadda.
30. P. Portaluppi, *Aedilitia*, vol. I (Milan, 1924), vol. II (Milan, 1930), in which all the power stations mentioned here are published, without text. Portaluppi moreover designed the architectural part of: the hydroelectric station of Molare for the Officine Elettriche Genovesi, 1920; the hydroelectric plant of Sottofrua for the Società dei Serbatoi Alpini, 1922; the transformation sub-station of Trecate for the Società Conti, 1922; the thermoelectric station of Piacenza for the Società Generale Elettrica dell'Adamello, 1925; in 1930 he participated in a competition for the hydroelectric plant of Ponte di Formazza for the Società Edison.
31. E. Conti, *Dal taccuino...* cit., pp. 61-2. When, in 1926, the Società Conti merged with Edison, the Verampio station was given the name of Ettore Conti, who since then used to sign as "Ettore Conti di Verampio."
32. *Ibid.*, p. 62.
33. Cf. *ibid.*, pp. 115, 146-8, and P. Portaluppi, *La Casa degli Atellani in Milano, per le nozze d'argento di S.E. il Senatore E. Conti e N.D. Gianna Casati* (Milan, 1922).
34. See for example eng. prof. G. Ganassini, "L'impianto idroelettrico..." cit.
35. See, for the technical description, Soc. Imprese Elettriche Conti, "Impianti in Val d'Ossola. Generalità. Impianto di Crego," excerpt from *L'Elettrotecnica*, vol. X, 5-15-25 January 1923.
36. See for the technical description Soc. Imprese Elettriche Conti, "Impianto di Valdo," excerpt from *L'Elettrotecnica*, 15 February-5 March 1925.
37. In *Aedilitia*, vol. I, p. 171.
38. The technical description of the plant is by G. Ganassini, "L'impianto idroelettrico..." cit.; the entire article is developed in three parts and more precisely: August 1926, pp. 636-55; September 1926, pp. 732-6; May 1927, pp. 491-508.
39. G. Minnucci, "L'architettura e l'estetica degli edifici industriali," in *Architettura e Arti Decorative*, V, 1926, pp. 481 ff.
40. *Ibid.*, p. 490.
41. Reference is made to C.E. Gadda's novel, "Un concerto di centoventi professori," (1934) in *L'Adalgisa. Disegni milanesi* (Milan, 1994), p. 171. It does not seem irrelevant (without however claiming to enter in the merits of a possible direct relationships between Gadda and Portaluppi) to emphasize that the name of the architect is recurrent in several texts by Gadda (attentive reader, among others, of the *Guerin Meschino* journal) while Piero Gadda Conti, nephew and then adoptive son of Ettore Conti, was also the cousin of the writer. On Gadda and the architects of the period see A. Grimoldi, "Il gaddesco architetto Basletta e altri architetti," in *Milano è una brutta e mal combinata città* (Milan: IN/Arch, 1994), pp. 39-51.

16

17

1

Jean-François Lejeune

Democratic Pyramids: The Works of The Tennessee Valley Authority

Long Shot (LS): Explosion of dynamite.
Medium Shot (MS): Steam shovel scooping tons of rock.
MS (Low Angle): Pneumatic drill operator with breathing mask.
LS (High Angle): Dam under construction from moving crane.
LS: Pouring concrete for dam.
LS: Controlled water flowing thunderously.
LS: Explosion of dynamite (as above).
MS: Pneumatic drill (as above).
LS: Explosion of dynamite (as above).
MS: Pneumatic drill (as above).
LS: Pouring concrete (as above).
LS: (With dissolve transition): Before/after shots of gorge now blocked by the Norris Dam.[1]

At the 1938 Venice International Film Festival, Pare Lorentz's *The River* was named best documentary and James Joyce observed for the press that the narration contained "the most beautiful prose I have heard in ten years."[2] Using "intellectual montage" in the manner of Sergeï Eisenstein, the film produced in 1937 by the US Farm Security Administration told the story of the Mississippi and Tennessee Valley. Its epic message reflected the New Deal's credo that, following decades of abuse, short-term exploitation, and mismanagement of natural resources, the problems of the Nation would not be helped by the rebuff of technology but the intelligent guidance of planners to coordinate its application.

Brainchild of the First Hundred Days of the Roosevelt Administration the Tennessee Valley Authority was established on May 18, 1933, "with the broadest duty of planning for the proper use, conservation, and development of the natural resources of the Tennessee River drainage basin and its adjoining territory for the general social and economic welfare of the Nation."[3] The condition of the valley was "an extreme example of a familiar US history: first the settlers pushed out the Indians, then when it was safe, the big planters squeezed out the settlers, and the lumbermen stripped the forest and the coal operators skimmed the cream of the mines."[4] As a result, poverty was widespread, and the socio-economic crisis made worse by the destructive violence of the yet untamed forces of water. While the initial objective of TVA's main sponsor, Senator George W. Norris, aimed at producing and distributing publicly controlled, low-cost hydroelectric power to farms, industries, and private households, the Acts of Congress went far beyond.[5] Under the direction of engineer Arthur E. Morgan, lawyer David E. Lilienthal, and agriculturist Harcourt A. Morgan, the Authority

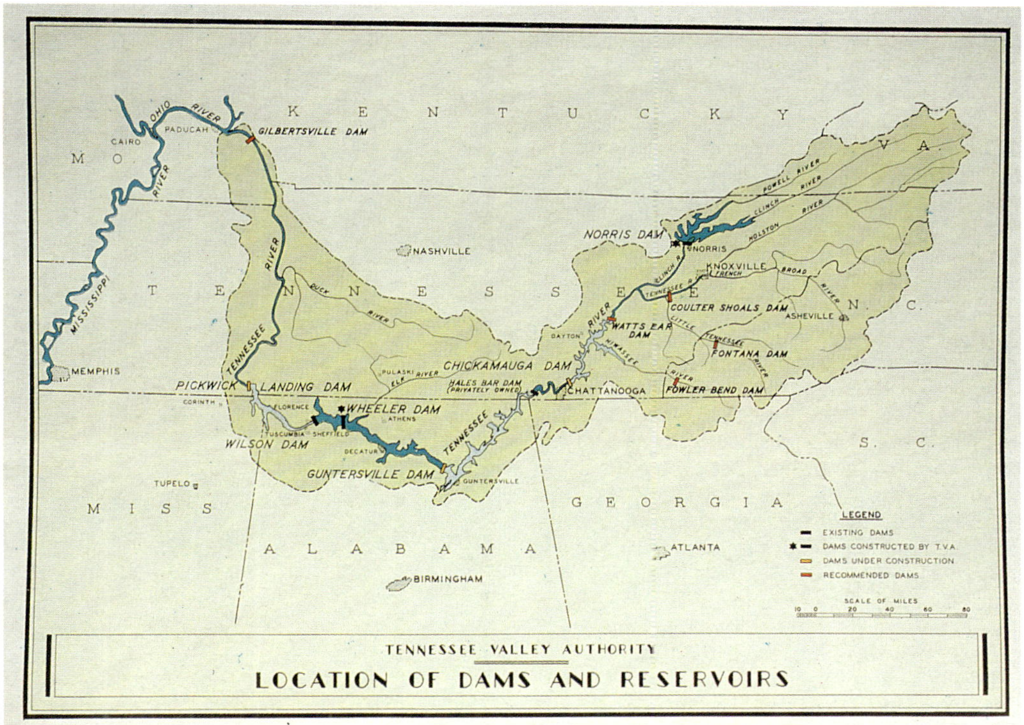

2

was empowered to control the floods through the construction of a unified system of dams and lakes, to make the river navigable for transportation, to reforest and restore fertility to the depleted fields, and to foster "studies, experiments, or demonstrations as may be necessary" to the orderly physical, economic, and social "unified development of [the forty-one thousand square miles] of the Tennessee River system."[6]

Hoover Dam and Gordon F. Kaufmann

Conceived under the conservative Hoover presidency (1928-32), Hoover Dam, the first and largest of the decade (1931-6), pioneered the policy and aesthetics of public works that the Tennessee Valley Authority would emulate and expand. Located in a desert area thirty miles from Las Vegas on the Colorado River, the dam was originally designed by the Bureau of Reclamation to rise 726 feet from bedrock as an arch-gravity structure; yet, its elegant shape conflicted with uninspired classical references. Hired to work in the new town of Boulder City, the London-born and Los Angeles architect Gordon B. Kaufmann (1888-1949) eventually redesigned the dam into a "masterpiece of Modernism" that belied – like the TVA works after it – Le Corbusier's assertion that "the great functional structures were simply the works of engineers."[7] Kaufmann streamlined the crest, spillways, and in-take towers, and converted the factory-like powerhouse into an abstract, monumental wall of concrete in front of which he put a series of box-like switch houses built on stilts and topped by electric transformers – a modernist scheme that recalled André Lurçat's early works. He also staged the competition for the monument to the dam won by the Norwegian sculptor, Oskar J.W. Hansen (1892-1971): two thirty-foot tall *Winged Figures of the Republic* were poured in bronze on top of a terrazzo platform decorated as a celestial map locating the astronomical time when President Roosevelt dedicated the dam on September 30, 1935. The economic and regional impact of Hoover Dam was formidable: on the short term, it created work for five thousand workers; on the long term it would not only spur the growth of California agriculture, but it eventually transformed Las Vegas into the "electric" recreation capital of the West.[8]

Robert A. Wank (1898-1970), first TVA architect

In a letter to the architectural historian Walter Creese, Earle Draper, TVA's first Direc-

On the preceding pages:
1. Ch. Sheeler, *Suspended Power*, 1939. Dallas Museum of Art.

2. Tennessee Valley Authority. Location of dams and reservoirs. From: *Report to the Congress of the Unified Development of the Tennessee River System*, submitted by the Board of Directors of the Tennessee Valley Authority (Knoxville, Tennessee, 1936).

tor of Regional Planning Studies, explained that he had selected Robert A. Wank as Chief Architect because he was "entirely original" and not "wedded to any style or tradition."[9] Born and educated in Hungary, Wank emigrated to the US in 1924. There he worked for the New York firm Fellheimer & Wagner for whom he designed the powerful and much admired "hymn to steam power," the Cincinnati Union Terminal (1929-33).[10]

At TVA, he was first hired to work on Norris Village, before being asked by Chairman Morgan to review the dam proposals. Wank was easily accepted by the team of designers, "dominated by engineers raised in the European tradition, Norwegians, Germans, and even Russians." His method was "to identify those parts of the project where the impact of architecture and the contribution of architectural design would concentrate most heavily. Thus he found in the design of generator halls a really significant opportunity to move a generation that had been told by Henry Adams about 'The Virgin and the Dynamo,' and led to rank it with the Chartres cathedral."[11]

Of the twelve dams included in the Unified Development Plan of 1936, three – Norris Dam (1933-6), Hiwassee Dam (1937-44) and Fontana Dam (1942-5) – were high, flood-storage, gravity-type structures located on the most dangerous tributaries of the Tennessee River.

Construction on Norris Dam started immediately on October 1st, 1933. With Gordon B. Kaufmann as consultant, and Albert Kahn as adviser in Detroit, Roland A. Wank and Chief engineer Theodore B. Parker were prompt to overhaul the Bureau of Reclamation's plans of 1927.[12] The project had none of the neoclassical, aqueduct-like elegance of Wilson Dam, its predecessor in the region: it was a monotonous design, with no relation to the landscape. Wank and Parker transformed the dam into a dynamic arrangement, where the guideways of the ski-jump spillway, the abstract, "metaphysical" control tower, and the vertical joints in the 260-foot high concrete wall counterbalanced the horizontal lines of the powerhouse and the thin, seemingly floating roadway supported by a "K-structure" bridge over the spillway.[13]

At Hiwassee, perhaps the most beautiful of the dams, the designers adopted a similar concept, but with an outdoors powerhouse built on a platform above the downstream side. The gracefulness of the spillway, the toy-like appearance of the gantry sliding on the roadway like a moving triumphal arch, the audaciously streamlined, bolted steel and alu-

3

4

3, 4. Norris Dam under construction
looking upstream in 1935.
Photos Rell Clements.

5. H. Ferriss, Boulder Dam Overlook, 1941.
Collection Avery Architectural Library, Columbia University, New York.

6. R.A. Wank (Tennessee Valley Authority-TVA), Norris Dam and powerhouse, first version, 1934. S.H. Gurnee renderer.

7. R.A. Wank (TVA), Hiwassee Dam, suggestion for downstream elevation, 1938. Drawn by M. Bianciulli.

8. Norris Dam and powerhouse from east abutment in 1937. Photo TVA.

minum powerhouse gantry at the bottom of the narrow gorge created a unique composition in the history of architecture, a lyrical and poetic encounter between machine, landscape, and raw concrete. Seen from the river, Norris and Hiwassee Dams rose as potent evocations of Antonio Sant'Elia's "Powerhouse" drawings of 1914. Inspired by German expressionist architecture, Hugh Ferriss' visions of "The Metropolis of Tomorrow," and perhaps by Tony Garnier's "Cité Industrielle," Wank even asserted having given the dams "an urban atmosphere, should a city [in the future] encompass the reservoir."[14]

The seven Tennessee River structures – Wheeler (1933-6), Pickwick Landing (1935-8), Guntersville (1935-9), Chickamauga (1935-40), Watts Bar (1939-42), Fort Loudoun (1940-3), Kentucky-Gilbertsville (1938-44) – as well as Cherokee (1940-2) and Douglas (1940-2) – were concrete and earth-fill, multiple-span dams, ranging from 2,960 to 8,422 feet in length and 72 to 206 in height.[15] Generally terminated at one end by the powerhouse and at the other by a navigation lock, they harnessed the river and regulated the nine-foot deep, 650-mile long, navigable channel from Knoxville to the Ohio River. Architecturally, each dam formed a unique axial composition based upon a rigorous constructive and typological system. The gantry cranes moving along the crests were a distinctive feature, inseparable from the architecture. Their design evolved along the years and gave an identity to each structure: the machinery of pipes, wheels and pulleys exposed at Wheeler Dam – what Hamlin called "the old romanticism of the machine" – disappeared from the streamlined versions (Pickwick, Chickamauga, Watts Bar) where all the equipment was housed in a plain, round-cornered protective box.[16]

At Kentucky, TVA photographers could not help but accentuating the expressionism of the machine.

Started in 1942 a month after the attack on Pearl Harbor and completed in December 1944, Fontana Dam was a model of pure and deprived rigor that reflected the new political conditions and specific technological choices. Located in a steep and narrow gorge, the highest of TVA's dams (480 feet) had no spillway, no elevator tower, or crane to counteract its horizontal lines: the overflow was channeled in two tunnels dug into the mountains, whose huge mouths were designed to project water 150 feet in the air and 400 feet downstream. With its five-story high powerhouse and its panoramic visitor center accessible by a cable

5

car, Fontana was a modern masterpiece that mirrored, willingly or not, Wank's growing attraction to the International Style.[17]

"Built for the People of the United States"

The dams were "Built for the People of the United States," a motto of pride and public ownership that was displayed prominently in every facility in large Déco-style aluminum letters, and helped "replenish the quivering American public aspirations."[18] Poured-in-place concrete, contrasted with windows and doors in stainless steel and aluminum, was the material of choice for the powerhouses and other structures that needed to resist high water floods. The alternating horizontal-vertical checkerboard pattern of formworks experimented at Norris was systematically used to reduce the impact of the long and high surfaces of concrete and diminish the effects of weathering. In some situations such as Chickamauga, the pattern was made more complex by inserting narrower panels of plywood to get more relief and bands of smooth surface between the board surface. The volumetric purity of the powerhouses and their adjacent buildings was accentuated by the thin, flat roofs, built without parapet walls, and covered with light slabs of concrete that made them attractive from higher viewpoints.

Inside, the powerhouses were conceived as modern temples, open to the people who were invited to attend, generally from an upper platform, the ceremonial and quasi-mystic rites of technology and electricity production: "[like in] early Christian basilicas, the very warmth of the surface materials, the veneers – marble, cork, chrome, terrazzo, and colored ceramic-tile – were calculated to mysteriously nourish the habits of denial and asceticism for the ever-present faith in technology."[19] The spectacular black and white photographs of E. Romska and E. Neukom do not make justice to the bold color scheme: in Pickwick's descriptions, "the generators are in dark green and white metal, tile wainscoting is in green, while the walls are lemon yellow and light green." In the Watts Bar turbines room, the walls were "of light-blue terra-cotta, the floors of gray ceramic tiles, and the machines yellow and tan."[20]

Surfaced with aluminum, chromium, travertine, and terrazzo, the reception rooms were more modest but equally impressive in their exaltation of transparency, fluidity of space, and "modern" horizontal lines: the best were at Chickamauga and at Guntersville whose grand double staircase echoed a New York film palace. Wank's quasi-obsessive attention to detail was carried on above and around the dams with the panoramic approach roads and parking lots, the elegant bridge guardrails, the Tennessee marble water fountains, the lamp-posts, and more notably the scenically located visitors buildings.

Proud to be an American

In 1941, nine years after the International Style Exhibition where public architecture and social implications had been notoriously absent, the Museum of Modern Art in New York dedicated an entire show to TVA architecture. For Lewis Mumford the dams represented "modern architecture at its mightiest and its best" and evoked "the largely unconscious precedents of our [American] grain elevators and storage warehouses and coalbins."[21] Other critics were hesitant to embrace such a richness of effects and textures. Afraid of discovering any signs of "decoration" they praised TVA's "puritanical" functionalism, contradicting Wank who had declared that "the design should reflect not merely the austerity of the present but also the greater affluence of the years to come."[22]

Undoubtedly TVA architecture was difficult to categorize: its size and the impact of engineering constraints set it definitely apart – with the exception of Fontana – from the reductive canons of the International Style. Similarly, and contrary to Walter Creese's postulate, it was neither "European monumentalist" nor Art

9. Hiwassee Dam and powerhouse, 1940.
Photo TVA.

6

7

Déco. Admittedly, memories of Otto Wagner's Nussdorf and Kaiserbad lock buildings in Vienna could be found, for instance, at Pickwick Dam and Junction, and Art Déco, streamlined influences were visible in many of the powerhouse interiors. Yet, renderings from 1934 show that the redesign of Norris Dam – which established the basis of the dams' aesthetic system – was a phased process and that Wank rejected a first Déco version of the powerhouse in favor of a more modern, horizontal design.[23]

Thus it is more accurate to say that Wank's architecture reflected the diversity of American modern architecture during the decade of "The Machine Age." Echoes of Norman Bel Geddes' projects, the Chicago Century of Progress Exposition of 1933, the Philadelphia Savings Fund Building (Howe & Lescaze, 1928-32), or the McGraw Hill Building in New York (Hood & Fouilhoux, 1930-1) could be found in many powerhouses. "Electric" buildings such as the General Electric distributors in Brooklyn and Queens (Hood, Goodley & Fouilhoux, 1929-33) displayed many details later applied at TVA. Classical Modernism was also influential, like at the Guntersville powerhouse whose general mass resembled Paul Cret's Central Heating and Refrigeration Plant in Washington DC (1933-4).[24]

Earle Draper, Norris, and Regional Planning

Dams were at the heart of the American modern psyche. Millions of visitors flocked to Norris, Hiwassee, and in the West, to Boulder and Grand Coulee. On November 23, 1936, the first cover of *LIFE* featured a full page photo of the machine/totem-like concrete gates of Fort Peck Dam, Montana. The century-old American fear of monumentality as symbol of autocratic religion and dictatorships had been won over by the "democratic pyramids," the American Valhalla of the New Deal. With their sheer volume, power, generosity, and apparent tranquillity, they merged into the wilderness while reinventing it, making it less awesome, more accessible, and more beautiful.

Under the direction of Earle S. Draper, planner and former associate of John Nolen, the landscape was remodeled, old roads relocated and new ones opened; in the first four years, extensive public recreation areas such as parks, beaches, and boat landings, were created around hundreds of square miles of new lakes. Yet, under constant attacks by the Republican forces dominating the Congress, most of TVA attempts to publicly control land along the shores for "regional planning" failed.

8

9

10

10. R.A. Wank, Watts Bar Control Building, 1940. Drawn by M. Bianciulli.

Among the three TVA directors, Arthur E. Morgan – biographer of Edward Bellamy, author of the utopian novel *Looking Backward 2000-1887* – was, until his dismissal in 1938, the "ideological head" of TVA. For Morgan and the group known as "Regionalists" – particularly the New York based Regional Plan Association of America – the use of electricity was to encourage communication, decentralization of population and industry, and eventually autonomy from the big industrial centers.[25] In creating TVA, Roosevelt was "heavily committed in principle to the RPAA line."[26] Benton MacKaye – an influential member of RPAA – saw in the Authority the long-hoped-for opportunity to implement his plan of 1921, "The Appalachian Trail: A Project of Regional Planning," for a linear city consisting of "townless highways" and "highwayless towns."

In 1922, Henry Ford had unsuccessfully planned a 75-mile long city for Muscle Shoals. MacKaye's project of 1933 integrated his New York Regional Plan of 1924 to create a two-thousand-mile narrow and long chain of cities between Tennessee and New York.[27] Designed by Draper as "a subsistence community embracing agricultural, industrial, and intellectual pursuits," Norris Village was conceived as an element of this regional plan.[28] With its romantic labyrinth of curving roads and open spaces, and its vernacular architecture, reminiscent of Tennessee cabins and dogtrots but modernized by "electric kitchens," Norris was exceedingly rural, and an unconvincing ideological remedy to the national urban crisis.

"America Needs You"

TVA's economic output was considerable: in 1941, TVA had more than 42,000 employees, many of them in construction. 800,000 farmers were involved in farm demonstration programs, tens of thousands in industrial development, particularly in aluminum production. In 1940, Talbot Hamlin wrote:

> TVA is the world's most striking contemporary example that large-scale planning is possible in a democracy; that no such false efficiency as that of dictatorship is necessary to produce great national works, conceived and executed for the benefit of all the people.[29]

Ironically, following Lilienthal's memorandum of "national emergency" of 1939, the Authority – originally created as "a moral force for World Peace" – was rapidly converted into a "weapon of war." Denouncing the slow response of Congress and the private sector, Lilienthal could proudly hold forth at the begin-

11

11. Watts Bar Dam, Control Building looking west from top of powerhouse in 1942.
Photo E.O. Romska.

12

12. Watts Bar Dam, Control Building Visitor Lobby looking east in 1943. Photo E.E. Neukom.

13, 14. Chickamauga Dam, exterior view of downstream face of powerhouse and powerhouse reception room looking northwest in 1940.
Photos E.E. Neukom & R.H. Burdick.

ning of the war that "the planes that are flying today and will be flying tomorrow would literally be impossible if it were not for the TVA."[30] The social experiments of the first years, like the TVA adult education program during the construction of Norris Dam, yielded to patriotic discipline and around-the-clock labor. On the planning front, the idyllic spirit epitomized at Norris vanished and the Fontana village was conceived as experimental site for the first grouping of prefabricated trailers in the country.[31]

The point of no return was reached when, in 1942, Oak Ridge, near Knoxville, became one of the secret sites of the Manhattan Project: most of the energy production was channeled to the production of enriched uranium for the construction of the atomic bomb.

At the end of the 1950s, one third of the abundant and inexpensive power generated by the dams and by the many ecologically harmful coal-steam plants built after the war, was squandered to build up the Cold War nuclear arsenal and supply the atomic power plants.

13

14

1. I have borrowed the expression "democratic pyramids" from L. Mumford, "The Sky Line: The Architecture of Power," *The New Yorker*, 7 June 1941, p. 58. The "script" is taken from: P.C. Rollins, "Ideology and Film Rhetoric: Three Documentaries of the New Deal Era (1936-1941)" in P. Rollins (ed.), *Hollywood as Historian: American Film in a Cultural Context* (Lexington: University Press of Kentucky, 1983), pp. 32-48. See also: L. Snyder, *Pare Lorentz and the Documentary Film* (Norman: University of Oklahoma Press, 1968).
2. P.C. Rollins, "Ideology and Film Rhetoric" cit., pp. 39-40. The quotation is in: W.L. White, "Pare Lorentz," *Scribner's*, January 1939, p. 9.
3. From: F.D. Roosevelt, Muscle Shoals Development, *Message from the President of the United States Transmitting a Request for Legislation to Create a Tennessee Valley Authority – a Corporation Clothed with the Power of Government but Possessed of the Flexibility and Initiative of a Private Enterprise* (Washington DC: US Government Printing Office, 1933), Message to 73rd Congress, 10 April, 1933. The literature on TVA is impressive and contains hundreds of volumes and articles. The most important are listed in D. Spaeth's bibliography, *The Tennessee Valley Authority: Architecture, and Planning, a Bibliography* (Ann Arbor: Vance Bibliographies, 1985), no. A.2196. The best work, and the only analysis of TVA to also discuss architecture and urban planning in great details, is by W.T. Creese, *TVA's Public Planning: the Vision, the Reality* (Knoxville: University of Tennessee Press, 1990). Also recommended is the exhibition catalog: M.

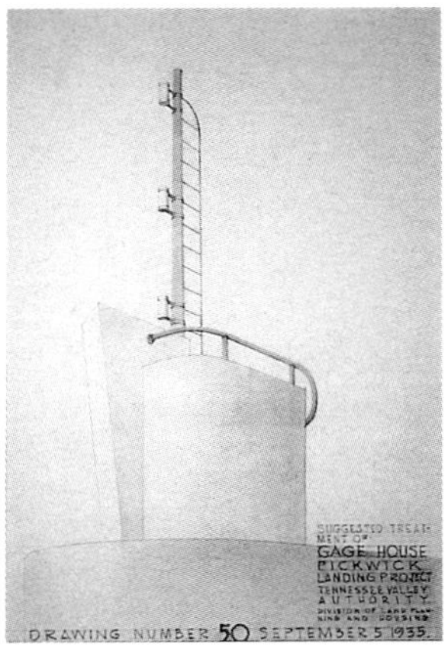

15

15, 16. R.A. Wank, Pickwick Landing project, suggested treatment of gauge house and proposed treatment of intersection of dam and lock, 1935. Drawn by M. Bianciulli.

Moffett and L. Wodehouse (eds.), *Built for the People of the United States: Fifty Years of TVA Architecture* (Knoxville: University of Tennessee, 1983). Other fundamental references are the government publications such as *TVA: Its Work and Accomplishments* (Washington DC: US Government Printing Office, 1940).

4. *TVA Program in Text and Pictures*, reprinted from *Architectural Forum*, August 1939 (Knoxville: TVA, 1939). About two million people lived in the valley in the 1930s. The amended TVA Act of 1935 helped to expedite TVA's supply of power.

5. The first step in controlling flooding and transportation on the Tennessee River went back to 1916, when the Army Corps of Engineers started construction of Wilson Dam to provide power for the military nitrate plants at Muscle Shoals. In 1921, Henry Ford offered to purchase the Muscle Shoal installation, but Senator George Norris (1861-1944), a progressive liberal Republican, opposed the sale. From then on, he became the first advocate of public control of electricity production and was the main artisan of the creation of the Tennessee Valley Authority. See P.J. Hubbard, *Origins of the TVA. The Muscle Shoals Controversy, 1920-1932* (Nashville: Vanderbilt University Press, 1961).

6. *Report to the Congress on the Unified Development of the Tennessee River System* (Knoxville), March 1936, pp. 9-10. Harcourt Morgan was dean of University of Tennessee's college of agriculture and a strong partisan of soil-building as a moral and scientific issue. David Lilienthal was a fast-track lawyer graduated from Harvard Law School in 1923. For A.E. Morgan, see his book, *The Making of TVA* (Buffalo: Prometheus Books, 1974).

7. From R.G. Wilson, "Machine-Age Iconography in the American West: The Design of Hoover Dam," *Pacific Historical Review 54*, November 1985, pp. 463-97. For a discussion of Kaufmann's career (1888-1949), see J.F. Muntz, "Gordon B. Kaufmann: California Classicism" and R.G. Wilson, "Gordon B. Kaufmann and Modernism," in *Johnson-Kaufmann-Coate: Partners in the California Style* (Claremont: Capra Press, 1992). Kaufmann also designed Grand Coulee Dam (1933-4) and Shasta Dam (1938-45).

8. The dam was so powerful that not until 1961 were all the power units put in service (R.G. Wilson, "Machine-Age Iconography in the American West" cit., p. 491). Hansen (1892-1971) was Norwegian.

9. Quoted from W.T. Creese, *TVA's Public Planning* cit., p. 147 (Earle Draper to Walter Creese, 16 September 1973). See F.A. Gutheim, "Roland Wank, 1898-1970," *Architectural Forum* 133, September 1970, pp. 58-9. Wank worked at TVA from 1933 to 1944. He was also chief architect of Greenhills (Ohio) and founded his own firm Wank, Adams & Slavin in 1945. He worked for the New Jersey Turnpike and was involved in industrialized construction (see R.A. Wank, "Integrated Design is a Joint Responsibility," *Pencil Points* 25, October 1944, pp. 48-53).

10. The terminal had an idiosyncratic half-circular roofline and centered on a rotunda 106 feet high and 180 feet in diameter. Its interior was decorated with glass mosaics by Austrian artist, Winold Reiss. For more discussion about the Central European immigration to the US, particularly from Austria, see M. Boeckl (ed.), *Visionäre & Vertriebene: Österreichische Spuren in der modernen amerikanischen Architektur*, exhibition catalog (Vienna: Ernst und Sohn Verlag, 1995).

11. F. Gutheim, "Roland Wank, 1898-1970," *Architectural Forum* 133, September 1970, pp. 58-59. H. Adams, *The Education of Henry Adams, an Autobiography* (Boston-New York: Houghton Mifflin, 1918), p. 380.

12. *Report to the Congress* cit. Asked to handle a dispute between engineers and architects in 1937 on the design of Guntersville Dam, Kahn became later a permanent consultant to the TVA.

13. For the full description of Norris Dam, see Tennessee Valley Authority, *The Norris Project, A Comprehensive Report on the Planning, Design, Construction, and Initial Operations of the Tennessee Valley Authority's First Water Control Project* (Washington DC: US Government Printing Office, 1941). Similar detailed reports exist for each TVA dam.

14. W.T. Creese, *TVA's Public Planning* cit., p. 174. The quote is from the interview of Roland Wank by Gilbert Stewart, Jr., TVA Information Office, 6 August 1965, TVA, 2.

15. The drop along the channel was about 520 feet.

16

17. R.A. Wank, Guntersville powerhouse, 1937.
18. Kentucky Dam powerhouse and gantry crane in 1940. Photo TVA.
19. R.A. Wank, Gilbertsville (Kentucky), preliminary study of control building, 1939. Drawn by M. Bianciulli.

17

18

19

Cherokee and Douglas, built on tributaries of the Tennessee, were built to produce more power for the war effort. The total number of turbines installed from 1933 to 1944 was 61 units.
16. T.F. Hamlin, "Architecture of the TVA," *Tennessee Valley Authority Architecture as Published in the Magazine Pencil Points*, reprinted from *Pencil Points*, November 1939, TVA, 1940.
17. W.T. Creese (*TVA's Public Planning* cit.) argues of war times and economies to explain Fontana's supposedly less appealing image. While his explanations are valid, I disagree with his assessment. See also note 24.
18. W.T. Creese, *TVA's Public Planning* cit., p. 152-3. Starting with Norris, Wank is generally credited for making the dams accessible to the public (W.T. Creese, *TVA's Public Planning* cit.; F.A. Gutheim, "Roland Wank" cit.; T.F. Hamlin, "Architecture of the TVA" cit.). According to Wilson, Gordon Kaufmann was the real "pioneer": the most spectacular interior was not at TVA but at Hoover Dam where he coordinated the interior works with Allen True.
19. W.T. Creese, "TVA as an Allegory," in M. Moffett and L. Wodehouse (eds.), *Built for the People of the United States* cit., p. 61.
20. W.T. Creese, *TVA's Public Planning* cit., chapter 4.
21. L. Mumford, "The Sky Line: The Architecture of Power" cit., pp. 58-60. He particularly praised Hiwassee and Guntersville dams but expressed regrets that the lack of space prevented the organizers "to indicate the architectural treatment of the whole landscape."
22. Quoted by W.T. Creese, *TVA's Public Planning* cit., p. 152: Roland Wank, interview by Gilbert Stewart, Jr., TVA Information Office, 6 August 1965, TVA, 1.
23. Let's add that Wank successfully vetoed the attempts by sculptor Jo Davidson to adorn Norris Dam's spillway with a large figure of Atlas (W.T. Creese, *TVA's Public Planning* cit., note 44, p. 177).
24. For these references, see R.G. Wilson, "The Machine in the Landscape," in *The Machine Age in America, 1918-1941* (New York: Harry N. Abrams, 1986). In his insistence on interpreting Wank's architecture as Déco and European, W. Creese makes dangerous generalizations. For him, in absence of "glass walls," the powerhouses are not "modern" (p. 160): a remark which ignores the constraints of water pressure in case of flooding. He also writes (p. 173) that "Art Déco tended to be weaker than it ought to have here for use in a democratic setting," an ambiguous and unfair statement in favor of modernism that contrasts with his criticism of Fontana.
25. E. Bellamy, *Looking Backward 2000-1887* (Boston: Houghton Mifflin, 1888). See also E.S. Draper, "Regional Planning and the Tennessee Valley Authority," Lecture at the Harvard Uni-

20

20. Fontana Dam.

21. Fontana Dam, powerhouse generator room in 1945. Photo E.E. Neukom & J.E. Fair.

versity School of City Planning, 14 January 1935; F. Dal Co, "From Parks to the Region: Progressive Ideology and the Reform of the American City," in *The American City: From the Civil War to the New Deal* (Cambridge Mass.: The MIT Press, 1979); M. Scott, *American City Planning Since 1890* (Berkeley: University of California Press, 1969); and J. Dahir, *Region Building: Community Development Lessons from the Tennessee Valley* (New York: Harper & Brothers, 1955).

26. P. Hall, *Cities of Tomorrow: An Intellectual History of Urban Planning and Design in the Twentieth Century* (New York: Basil Blackwell, 1988).

27. *The Appalachian Trail*, Journal of the American Institute of Architects, 1921. Tracy B. Augur, Draper's assistant for Norris Village, was also a member of RPAA. The forester Benton MacKaye (1879-1975) was strongly influenced by Patrick Geddes and Lewis Mumford. On the rural myth, see for instance: G. Ciucci, "The City in Agrarian Ideology and Frank Lloyd Wright," in *The American City: From the Civil War to the New Deal* cit., and the important work by L. Marx, *The Machine in the Garden* (New York: Oxford University Press, 1964).

28. *Built for the People of the United States* cit., p. 10. See the article "The TVA Town of Norris Will Soon Be Built. No Run-Down Outskirts Will Ever Mar It; Nor Will Through Traffic Disturb It," in *Architectural Forum* 60, January 1934: 77-78. On Muscle Shoals, see note 5. On Earle Draper, see J.L. Hancock, *John Nolen and the American City Planning Movement, a History of Culture Change and Community Response, 1900-1940* (University of Pennsylvania, 1964); and also M. Crawford, "Earle S. Draper and the Company Town in the American South," in *The Company Town, Architecture and Society in the Industrial Age* (New York: Oxford University Press, 1992). Norris' experiment was not repeated: workers' communities at Hiwassee, Wheeler, Pickwick, were mere transitory camps, but well equipped and designed with great care. Although it has been systematically ignored by historians, Boulder City, the new town planned in 1931 near Hoover Dam by the Department of the Interior, anticipated Norris Village and pioneered the concept of a permanent city linked to the dam. Designed by planner S.R. De Boer, it anticipated its plan and its Spanish-inspired typologies were much more convincing than at its Tennessee counterpart.

29. T.F. Hamlin, "Architecture of the TVA" cit., p. 45.

30. D.E. Lilienthal, *TVA and the War; A Message to Employees at the End of TVA's First Year of War* (Washington DC: US Government Printing Office, 1941). In taking control of the Muscle Shoals complex in 1933, TVA had in fact accepted national defense as one of its responsibilities. By

21

22. Fontana Dam, display board of progress photographs from January 1942 to December 1944. Photo E.E. Neukom & R.R. Lewis.

accelerating the construction of dams, more power was available at the aluminum plants when the military build up started in 1942.

31. See also J.M. Chancellor, *The Library in the TVA Adult Education Program* (Chicago: American Library Association, 1937). The elegant "International Style" E-2 modular houses by S.H. Gurnee were apparently never produced (M. Moffett and L. Wodehouse, eds., *Built for the People of the United States* cit.). Interestingly, many TVA trailers were later shipped to England to provide temporary housing after the war damages.

I am greatly indebted to Robert E. Cole, architect at the Tennessee Valley Authority in Knoxville, Tennessee, who helped me select and obtain the photographic material from the voluminous archives of TVA. I also thank Anita Gross, registrar of Works on Paper, Marie Stewart, her assistant to the Registrar, and Jim Findlay, librarian, for their help at the Wolfsonian Foundation, Miami Beach. Special thanks to Hao Shan, graduate student at the University of Miami, and Roselyne Pirson, journalist and Ph.D. candidate at the University of Miami, who edited this paper and attracted my attention to the film *The River* and the quoted essay.

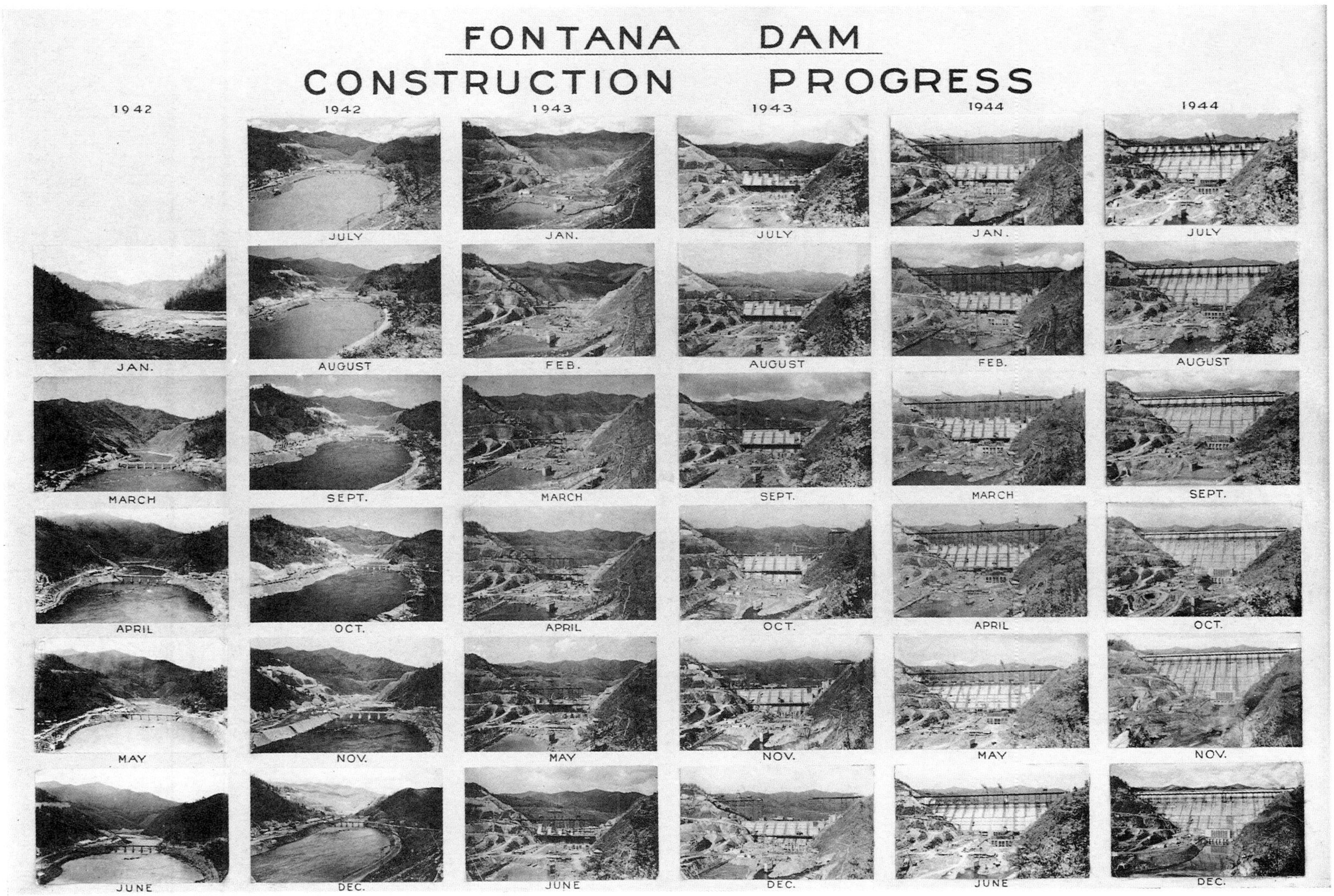

22

Alessandro De Magistris

Lights of the Revolution. Russian Electrification, Soviet Electrification

In December 1920 Engineer Gleb Krhizhanovsky activated an electric-light map of Russia to convince the Congress of Soviets (the eighth) to launch a general electrification plan. This has gone down in history with the acronym of GOELRO,[1] one of the most famous, if not the most famous, of the many that were launched in the Soviet period, with the proliferation of administrative institutions. During the course of 10-15 years, the plan envisaged the construction of 20 thermal power plants and 10 hydropower plants, the first of which, "Kashirskaya", began operating in 1922.[2]

In the precarious conditions in which the country found itself, the cost of that feat caused a momentary interruption in the provision of energy to part of Moscow, whose inhabitants, accustomed to harsh living conditions, hardly noticed what had happened. Yet the result was historically decisive, as is known, and opened one of the fundamental chapters in the history of the USSR, the centrality of which it is hard to overestimate. And this not only – obviously – in the understanding of the country's economic scenarios and strategies of development, but also as a specific monument for reflection on and the expression of the land theme and of architectural research.[3]

Actually, electrification had sound premises in Russia, thanks to the decisive presence of western companies and technicians, particularly the Germans, and to the key research and experimentation carried out in the army by the GAU (Glavnoe Artillerijskoe Upravlenie: Central Artillery Command). For some time – since the last twenty years of the 19th century – the new source of energy, with its applications, was transforming the urban environment, culture and psychology in Russia too. In the early 1880s the electrotechnical detachment (VI section) at the Imperial Russian Technical Association (IRTO: Imperial'skoe Russkoe Tekhnicheskoe Obshchestvo), the most authoritative seat of representation of the techno-scientific *intelligentsia*.[4] Shortly after that, the first filament bulbs were lit up, and began to replace the traditional gas or kerosene lamps in the streets

1

2

1. V. Gourkov, poster *Vo Imja Kommunizma* (In the Name of Communism), 1951.

2. *Give the Sun to the Night*, poster of the 1920s.
From: *Serdcem slushaja revoljuciju* (Leningrad, 1977).

3. N.G. Kotov, *Kapital'noe Stroitel'stvo*, poster of the early 1930s.
From: *Agitacija za shchast'e* (Bremen: Edition Temmen, 1994).

of some of the main Russian cities. In St. Petersburg, in 1882, the Cathedral of St. Isaac was lit using the new system.[5] Decisive steps, even if timid ones. Since here too, as in all the other grounds of modernization (from the telephone to the telegraph, from the new means of transportation to the other technical services that were to guarantee the city higher standards of hygiene and health),[6] the country continued to observe a considerable delay, caustically denounced by the specialistic press and by the accounts of western visitors, often betraying their ill-concealed disdain.[7]

There was nothing substantially unexpected in the proposal to launch a program of large-scale electrification, destined to innervate the main areas of the country, centralized and arranged starting from the identification of a series of economic regions.[8] Neither could the GOELRO, as such, call itself, as so many other programs in the first years of the revolution, a specific product of the first, painful years of the Bolshevik "revolution", as suggested by the official historiography. Many of its elements had been moulded for some time now on the Russian forge, and were to take concrete shape, perhaps under another name, whatever the outcome of the postwar period might have been. It is true, however, that these premises, fitting within the framework of revolutionary events, gave rise to an original mix, not so much from the standpoint of the specific choices, already inscribed in the layout and in the emerging features of Russian industrialization as well as in the conceptions and studies carried out before and during the war by the specialists, as from the implications and extent of the cultural and ideological reverberations, summed up in the famous Lenin aphorism: "Communism is the power of the Soviet plus the electrification of the whole country." This took on the features of a full-blown "technological utopia" in the meaning given by Howard P. Segal.[9] One of the many utopias, differently oriented and sometimes in conflict, running through the country in those tragic and decisive days.[10] Goals and research that had been under way since the early years of this century, as recent studies have shown, and – as elsewhere, moreover – had known a powerful acceleration with the nearing of the conflict, thus found a detonator and an amplifier for the new expression of already mature claims, borne and filtered by an élite, that of the "electrical engineers" or "electricians", who moved almost entirely to the new revolutionary shores (except for their being physically decimated by the harshness of the living conditions) – consisting

3

of specialized technicians gathered together in the offices and laboratories of St. Petersburg and Moscow; an élite particularly active in claiming its centrality in the events of the country. This is an unusual chapter, of a history of Russian and Soviet bureaucracy that precisely in recent years has been enriching itself with important elements and which can add a great deal to the understanding of the forms of Soviet land-bound thought.[11]

Electrification in the Architectural and Land Culture of the Early 1920s

In his recent important contribution, dealing with the political, economic and institutional expressions and the strategies of the new technical élites of the early Soviet-Russian electrification, J. Coopersmith[12] indicates 1926 as the limit of a periodization of the process in the post-revolutionary phase; beyond this threshold, we must speak of a new paradigm; changing from a driving, authoritative element, the program becomes a subordinate factor in the industrialization that was part of the "construction" of the five-year plan: with 1925-6, the author maintains, one can no longer speak of an electrification "in Russia," but of a "Soviet electrification."

This picture seems of key importance in understanding the meaning and the nuances that *elektrifikacija* had with Soviet architectural and urban planning culture, and the specific weight that the engineering and economic "fact" bears in supplying and directing, with Soviet-Russian disciplinary culture, an original vision of the land and social project, with its different projections and representations.

In fact, it is true that the early years of the new régime appear indissolubly linked with the terms of research linked to the Howardian model, and the timely welcome that the new ideas had with progressive intellectuals ever since the early years of the 20th century, especially thanks to the mediation of the German Gartenstadtgesellschaft, then equally doubtless appears the meaning that the developments of the program and the culture of electrification had in reshaping the contents of the model of the garden city, as this had been assimilated into the Russian technical and intellectual circles operating between the capital, Moscow, Warsaw, Riga and Charkov, in contact with the international circles. It does not seem rash to state that precisely the new institutional size of "Soviet" electrification can be linked to the very vitality of the Howardian hypothesis, "explored" in the early years of the revolution, with all its possible values and land scales, before entering a more complex circuit of mediations that flew into the "radical urban planning" of the first *pjatiletka*.[13] Along with the realization of the "workers' villages," directly linked to the new power plants,[14] of particular interest is the analysis – conducted by an extensive literature – of the first "Soviet" projects and urban planning schemes, always taking on a "land" scope, such as that developed by B.V. Sakulin within the VSNCh – already ready in 1918 – which was none other than the land concretization, in a decentralizing vein, of the programs of energy organization in the Moscow region.[15]

Proving to be more controversial, for this first period, was the reflection on the architecture of the power plants (thermal – TES, or hydroelectric – GES). Far from being considered marginal, until the mid-1920s, the design outcomes of these early experiences are probably more interesting inasmuch as they attest to the strong-rooted nature of the prewar lines, and of the capacity of the codes shared by the mature generation of architects who "entered" the revolution, to measure themselves against the large scale, and to establish a dialogue with the landscape, than for innovative elements as such. That which be-

4

4. I. Zolthovskij, G. Gol'c, MOGES, Moscow, 1927.

5, 6. Volkhovskaja GES, 1922-6. Front looking downstream, and inside view.

7. F. Shekhtel, project of the hydro-electric powerplant on the Dnepr, 1924.

gan to unfold and show itself again, in those years, had impeccably urban roots and motivations, and more than celebrating the magic word of the state control on the economy, of the great government order, preferred the channels and the more agile and dynamic lexicon, certainly ephemeral but also ideologically "ambiguous", rather than the semi-public commercial trusts that were appearing on the service-sector scene, or that of the trade unions, whose presence was observed by the workers' clubs. The earliest projects and the first creations, at least up to the mid-1920s – such as the Utkina Zavod' in Petrograd (1922), the Kizhilovskaja (1924), Nizhegorodskaja (1925) – also because of the marginal role played, in the project dialogue, by the architectural culture, did not constitute a privileged place of research; or if they were, they suffered from the limited degree of domesticity which the architects focused on the tasks assigned them.[16] With the exception, perhaps, of the Shaturskaja GES (1925) and the MOGES by Zholtovskij (which can still be considered as already participating in a subsequent phase),[17] a belated application of well-tested monumental motifs already put in focus, on an urban scale, in the variegated frame of a "Modern" which in the years before the war was declining towards forms of Classicism and Rationalism. This proves particularly clear in the handling of the openings.

Among the examples, mention can be made of the Zemo-Avchal'skaja GES (by E. Kal'gin and M. Machavariani) near Tbilisi and, above all, of the Volkhovskaja GES (56 thousand kW of power), built, with great difficulty – thanks to the decisive contribution of the Swedish company ASEA (Allmanna Svenska Elektriska) – between 1918 and 1926[18] following the project by Engineer Graftio and the architectural consultancy of O. Munc, V. Pokrovskij, A. Tikhomir, N. Gundobin; this was characterized in the lower tract of the imposing sequence of the 16-meter-high windowed arcades – which revealed the internal arrangement of the turbines[19] – formally tied to well-known formal motifs and accents, since the first decade of the century, in the large urban operations which had as their object the railway stations or the large commercial buildings.

The generational key, as well as the professional dialogue between specialists, appears important in this case too, thinking of the role that the architects already active in the flourishing "silver age" of the 1910s had, with respect to these demanding state orders. Paradigmatic of this is the project for the Dnepr

5

6

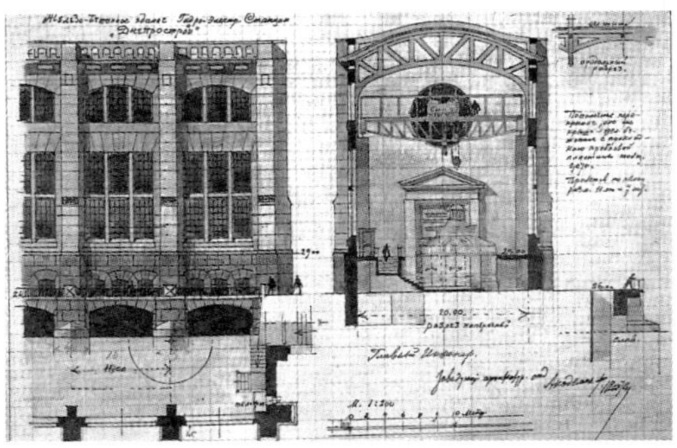

7

power plant and relative installation (Elektrogorod)[20] – the last one – developed by Feder Shekhtel (one of the few developed in the Bolshevik years that marked its premature human and professional decline), for the power plant on the Dnepr whose construction figured among the priority indications of the GOELRO.[21]

The Dneprovskaja GES

The construction of the power plant on the Dnepr – built in 1927-32 near Zaparozh'e and designed, with an extensive contribution by American specialists, by Prof. I. Aleksandrov, the academicians A. Vinter and B. Vedeneev and, as regards the architectural side, by V. Vesnin, N. Kolli, G. Orlov and S. Andreevskij – is representative of the fast modification taking place within the horizons of electrification in the late 1920s and the start of the first five-year plan.[22] The heritage of the GOELRO had by now turned into the Gosplan, and the term *energetika* remained a key word, relating to perspectives of very different scope compared with the initial phases of the revolution. The intervention was by now a part of a strategy of overall transformation of the land, whose energy and technological factors – though loaded with an autonomous propaganda burden – were considered in terms that were substantially instrumental with respect to the whole series of interventions envisaged, which were to affect the country's landscape, housing structure and productive organization.[23] Even if identified as a priority project by the GOELRO, at the time when the Dneprostroj builder's yard was set up, it in fact summarized and brought out the productive, social and land themes that were proper to the initial phase of the forced industrialization, expressing all the keys – including the symbolic ones – that were part of the "strategy" of the *pjatiletka*.

The Dnepr dam, which began operating in 1932, meant the erection of Europe's main hydroelectric power plant; affirming the solidity and dynamism of the new social and economic "order" of the "triumphant" Socialism period, but also crowning a historic ambition of Russian power – almost sanctioning a long-lasting legitimization – which originated at the end of the 18th century: that of making the Dnepr river navigable (it was hindered in various tracts) from the Black Sea to the Baltic Sea through a lock consisting of three basins, each of which 120-meter long and 18 wide.[24]

But it also meant the possibility of supplying with generated energy a whole economic region that touched the mines of the Donbass and all the new plants for the production of aluminum, high-quality steel and iron alloys, constituting one of the largest Soviet *kombinaty* in the sector of the means of production. Hence the complexity of a many-faceted intervention, which, along with the elements concerning the operation of the hydroelectric basin, implied the realization of a new and highly structured system of road infrastructures connecting the Ukraine with the regions of Central Russia. Moreover, the construction of a large Socialist urban housing development was planned, to unify the village of Dnepropetrovsk to the north of the dam with the old Zaparozh'e (Aleksandrovsk) on the southern bank.[25] Precisely in the context of the new neighborhoods of Zaparozh'e, among other things, one of the most interesting episodes in

8

8, 9. Pages from the issue of the *SA* magazine on Dneprostroj, 1929.

10, 11. The Dneprostroj.

Soviet architecture and urban planning of the 1930s took shape; in it it is possible to trace, within the space of a decade, many of the trajectories, the uncertainties and the mediations dominating the practice and theory of design.

Finally, there was a basic technical-political implication, of particular scope at that particular juncture, emerging from the use of the knowledge and experiences of technicians and thousands of foreign workers.

Western contributions were certainly no novelty in those years, which still seemed to promise a relative opening-up of the country,[26] but here for the first time they took on extraordinary proportions, embracing the supervision of the project (Cooper, Hugh L. & Company and Siemens AG), the construction of the turbines (New Shipbuilding, Drydock Company and General Electric), most of the materials and the equipment used on the worksite.[27]

From the standpoint of the architectural approach, the comparison between the primitive Shekhtelian project and the competition projects[28] gives both the measure of the radical change that had taken place on the operative level and the new perspectives of the theoretical debate. The project selected prevailed over those proposed by some of the country's better design groups – among which the two versions submitted by I. Zholtovskij[29] – crystallized in an exemplary fashion the themes and the keys of the constructivist dictate located on that "landscapistic" scale; by this stage the latter represented, in the years of the initial industrialization, the appropriate reference for action in terms of a space of which the city was no more than a residual element.

Sitting on the granite banks of the Dnepr, and standing in a flat, desolate landscape, the complex was dominated by the gigantic arch of the dam barring the water course, with its rhythm marked by the sequence of pylons. Beside this, the housing for the machines of the power plant (250-m long by 50-m wide) completed the intervention arranging a sort of counterpoint dictated by the horizontal sign of the long and narrow opening. The *Sovremennaja Architektura* magazine devoted a whole issue to this work, No. 6 of 1929, publishing site photos and texts, commenting on the main points of the proposal-manifesto. They thus brought out the possible levels of interpretation which were meant to sum up an era, that of the triumph of the five-year plan, and of its productivistic and technological myths; and, above all, of the "credo" (functional-constructivistic) that seamed to give concrete shape to it without any possible margin for discussion.[30]

9

10

11

Particular emphasis was given to the strictly functional justification of each individual project decision, such as the adoption of the "fenêtre en longueur" which characterized the power plant, brought back to the southern orientation of the main façade, and the need to reduce the exposure to the rays of the sun through the use of an overhang, thus improving the working conditions during the day.

The Years of the War and the Early Postwar Period

The years of the war, among other things, brought about the destruction of the Vesnin Dneproges; it was reactivated between 1947 and 1950, through the "philological" reconstruction of the parts destroyed and the enlargement of the hydroelectric center.[31] Yet they did not completely interrupt the electrification program, which was extended along the northern path of the Volga, thanks to the construction of the GESs of Uglichskaja and Rybinskaja, the fundamental pieces in the energy provision to the capital in the war years and in the restructuring of water communications between the Moscow region, the Volga, the Baltic Sea, the White Sea and the Caspian Sea.[32] A series of technical innovations were introduced in the planning of the northern power plants, such as the use of large prefabricated concrete covering panels, which allowed an initial rationalization of building techniques and processes, applied extensively in the construction of the large urban *kvartali*. A whole series of documents attest to the development of a debate of a certain interest, hinging upon the mediation between requirements of standardization (imposed by the scale of the intervention), and "artistic" imperatives[33] which were inextricably tied to the task, albeit a symbolic one, that these realizations were called upon to perform.

If the tones of a barely monumentalized rationalism shone through in the minor constructions, in the realizations which characterized the nodal points of the great economic design of reconstruction of the Soviet territory, the formal motifs dictated by the many possible local interpretations of a Classicistic historicism prevailed; the general aura of "Socialist realism" had far from defined and stabilized these motifs. They were the master lines of the large-scale operations of the early period of the reconstruction, which, in the early postwar years, were to delve deeper into the redesigning of the "Soviet" landscape between the Black Sea and the Caspian Sea.

The culminating realization of this program, paradigmatic of the strategies and real economies of the late phase of Stalin's totalitarianism, was the Cimljanskij hydoelectric complex (*gidrouzel*), designed by a collective headed by A. Poljakov; it was composed of a 13-kilometer dam, two locks,[34] a hydroelectric power plant, various service buidlings, and two housing developments, located on the right and left banks of the river, for the workers involved in the various navigation and energy production activities. As part of the Volga-Don water corridor, the center reproposed the triumphal cadences of these, aimed at celebrating victory in the "great patriotic war," through explicitly taking up the motifs of the Russian Neo-Classicistic tradition that developed between the 18th and the 19th centuries, achieving a balance between the different parts so as to obey the principle of the "overall design."[35] A composition which could be grasped not so much at close range as from an ideal (and dynamic) aerial standpoint from the side of the water basin, where the contrasting and tense interplay of the horizontal lines and projections stood out (towers, triumphal arches and the sculptural elements themselves). The symbolic and ideological values, generated by the very scale and land logic of the intervention – inscribed, as all the architectural-urban planning achievements of the period in an explicit "cosmological" vision – were summed up in the painting by D.A. Naldabjan entitled *Dlja shchast'ja naroda* (For the People's Happiness), the reproduction of which formed the opening pages of the relaunched *Architektura SSSR* magazine devoted to its presentation. The reader could admire Stalin, in the guise of "Generalísimo", and in the equivalent one of "Architect of Communism," surrounded by the party leaders, intent on arranging, or correcting the fundamental lines of the project on a plan. A simple point, here as in many propaganda manifestoes celebrating the completion of the country's main hydroelectric basins, of a design, by now substantiated from the geopolitical standpoint, which extending beyond the boundaries of the country, affirmed itself symbolically in the Moscow "tall buildings." Nothing in this iconography recalled the accents, always focused on the size of the work and the idea of the realization of a new productive process, that were prevalent until the 1930s in the celebration of similar enterprises. With all this, the honors of the work revealed themselves to be very ephemeral.

With the turning point that occurred precisely in the years of its construction, on a par with the other architectural symbols of the late Stalinian phase, the work's good fortune disintegrated very quickly. What was one of the most important sites of those years, linked to the presence of that parallel economy of forced labor that was – paradoxically – itself entering a crisis, was subjected to harsh criticism in the debates that reached the second Conference of architects held in Moscow in December 1955. This – just as the famous Khrushchev speech – was in fact none other than the mark of a tension animating the whole reconstruction phase, and is restored by the written accounts of the meetings held at the disciplinary institutions and crystallized in the comment made on the project by A. Fisenko.[36] Here the terms of criticism were still within the dictates of academic composition and Socialist Realism; the specific debt was the absence of a "character" of the complex ("[...] only the presence of the dam and the water tell us that this is the building of a hydroelectric power plant"). Three years from then, the judgement referred as a whole to the Classicistic references now being completely eliminated. New horizons were unfolding which led to the rehabilitation of the Constructivistic dictate.[37] The example to be followed, to which Poljakov was supposed to keep, was that of the Vesninian power plant on the Dnepr.[38]

12

13

14

12. Uglichskaja GES. Power station and dam.

13, 14. Rybinskij Gidrouzel, 1930s-1950s. Details of the dam viaduct.

1. Gosudarstvennaja Komissija po elektrifikacii Rossii (State Commission for the Electrification of Russia).
2. About GOELRO: RGAE, F. 5208, kat. 2, op. 1, d. 78; Plan elektrifikacii RSFSR. Doklad VIII s'ezdu Sovetov Gosudarstvennoj Komissii po elektrifikacii Rossii, izd. 2 (Moscow, 1955); H. Chambre, *L'aménagement du territoire en URSS* (Paris, 1959); G. Guroff, *The State and the Industry in Russian Economic Thought 1909-1914*, Ph.D. thesis, Princeton University, 1970; A.G. Cummings, *The Road to NEP. The State Commission for the Electrification of Russia (GOELRO): A Study in Technology, Mobilization and Economic Policy*, Ph.D. thesis, University of Maryland, 1988; J. Coopersmith, *The Electrification of Russia 1880-1926* (Ithaca-London, 1992).
3. In general: I.N. Magidin (pod redakciej), *Architektura gidroelektrostancii* (Moscow, 1954); A.K. Aleksandrov, *Ocherki po istorii energeticheskoj techniki SSSR. Proektirovanie gidroelektrostancij* (Moscow, 1955); D.G. Zimerin, *Istorija elektrifikacii SSSR* (Moscow, 1962).
4. The IRTO was founded in 1866. Inside it, in the late 1870s, an electrotechnical detachment was formed, with its headquarters in St. Petersburg and Moscow, institutionalized in 1880 with a Foundation Agreement of the 6th Section, at which 56 people participated.
5. *Elektrichestvo*, Nos. 10-11, 1889, pp. 125-6.
6. On the consequences and social images of the new technologies in the 19th century, see: D.E. Nye, *Electrifying America. Social Meanings of a New Technology, 1880-1940* (Cambridge-Mass.: MIT Press, 1990); D.E. Nye, *American Technology Sublime* (Cambridge-Mass.: MIT Press, 1994); C. Marvin, *Quando le vecchie tecnologie erano nuove. Elettricità e comunicazione a fine Ottocento* (Turin, 1994).
7. Cf. J.H. Bater, "The Legacy of Autocracy: Environmental Quality in St. Petersburg," in R.A. French, F.E.I. Hamilton (eds.), *The Socialist City: Spatial Structure and Urban Policy* (New York: Chichester, 1979); D. Browner, *The Russian City between Tradition and Modernity 1850-1900* (Oxford: Berkley, 1990).
8. The first power plant of considerable size, capable of a production of 57 thousand kW, was the

15

15. S. Buzjukov et al., standard project of thermoelectric power station, 1950.

16-18. Cimljanskaja GES, end 1940s-early 1950s. Different proposals. From: *Architektura CCCP*, No. 5, 1990.

"Elektroperedacha", inaugurated in 1914 at Bogorodsk, near Moscow.
9. H.P. Segal, *Technological Utopianism in American Culture* (Chicago, 1985), pp. 18-19.
10. R. Stites, *Revolutionary Dreams. Utopian Vision and Experimental Life in the Russian Revolution* (New York-Oxford, 1989).
11. *Cahiers du Monde Russe et Sovietique*, No. 4, 1991 (Spécialistes, bureaucratie et administration); A. Graziosi, *Stato e industria in Unione Sovietica. 1917-1953* (Rome, 1993).
12. J. Coopersmith, *The Electrification of Russia* cit.
13. V. Chazanova, *Sovetskaja architektura pervoj pjatiletki* (Moscow, 1980); A. De Magistris, "Culture architecturale et projet urbain dans les années 30," *Cahiers du Monde Russe et Sovietique*, No. 4, 1991, pp. 609-26.
14. V. Chazanova, *Sovetskaja architektura pervych let Oktjabrja* (Moscow, 1970).
15. *Iz istorii sovetskoj architektury 1917-1925 gg. Dokumenty i materialy* (Moscow, 1963); M. Astaf'eva-Dlugach, "Pervye schemy sociasticheskogo rasselenija v SSSR," *Architektura SSSR*, No. 6, 1970, pp. 14-17.
16. V.A. Kovalev, "Architektura teplovych elektrostancij, sooruzhavshichsja po planu GOELRO," in A.A. Strigalev (pod redakcej), *Problemy istorii sovetskoj architektury* (Moscow, 1980), pp. 15-23.
17. Today GES No. 1 im. P.G. Smidivica. Cf. *P Ezegodnik MAO*, No. 5, 1928. GNIMA P1a 8534/1; P1a 8534/3-6.
18. A.C. Sutton, *Western Technology and Soviet Economic Development* (Stanford, 1968); J. Coopersmith, *The Electrification of Russia* cit., pp. 148-9.
19. V.A. Myslin, "Architektura proizvodstvennych cechov," in *Problemy architektury. Sbornik materialov pod redakceij A.Ja. Aleksandrova* (Moscow, 1936), p. 70; V. Chazanova, *Sovetskaja architektura* cit.; T. Suzdaleva, "Otkrytie nasledija revoljucionnogo romantizma," *Architektura SSSR*, No. 2, 1989, p. 103; J. Coopersmith, *The Electrification of Russia* cit., pp. 148-9.
20. GNIMA im. Shchuseva, PI -11630/1-10. (Al'bom "Dneprostroj"). For the project (ink, pencil and watercolor on paper, 25 x 38.5 cm) GNIMA-PI-7034.
21. As is shown on the documents, the project was developed by F. Shekhtel, who was working within the Gosudarstvennoe Dneprovskoe stroitel'stvo (as the organization for designing the Dneprostroj was called) between December 1924 and March 1925. At the basis of the study was the general plan for the complex developed by Prof. I.G. Aleksandrov.
22. Linked to this phase of electrification is, among others, one of the most significant chapters of Italo-Soviet cooperation, linked to the name of Eng. Angelo Omodeo and his assistants, who worked for the Narkomzem and the Energocentr. Cf. M. Cecchini, "Due missioni tecniche italiane in URSS 1930-1931," *Storia Contemporanea*, No. 4, 1987.

16

17

18

23. V.A. Myslin, "Architektura proizvodstvennych cechov" cit., p. 76.
24. M. De Michelis, P.Pasini, *La città sovietica 1925-1937* (Padua, 1976).
25. G. Orlov, V. Lavrov, "Bol'shoe Zaparozh'e," *Architektura SSSR*, No. 3-4, 1933. G. Orlov, V. Lavrov, "Bol'shoe Zaparozh'e," *Architektura SSSR*, No. 12, 1939. T. Bazanova, "Drama v dvuch dejstvijach s prologom i epilogom," *Architektura SSSR*, 1991.
26. F. Bucci, *L'architetto di Ford, Albert Kahn e il progetto della fabbrica moderna* (Milan, 1992); A. Graziosi, *Stato e Industria in Unione Sovietica 1917-1953* (Rome, 1993); A. De Magistris, "Pervyj Podshipnikovyj Zavod. Il 'caso' della prima fabbrica statale di cuscinetti sovietica," *Le Culture della Tecnica*, No. 2, 1994, pp. 7-22.
27. A.C. Sutton, *Western Technology and Soviet Economic Development* cit.
28. G. Orlov, "Architektura Dneprovskoj gidroelektrostancii imeni V.I. Lenina," in I.N. Magidin (pod redakciej), *Architektura gidroelektrostancii* cit., pp. 47-8; E. Matveev, "Dneproges", *Architektura SSSR*, No. 12, 1970, pp. 7-10; M. De Michelis, P. Pasini, *La città sovietica 1925-1937* cit., p. 58.
29. Of the two projects presented by I.V. Zholtovskij, the first, the basic one, envisaged a stone-clad façade. A sequence of openings underlined by the ashlars opened onto the central balcony supported by the powerful corbels. In the marked Classicistic accent it presented simple, severe forms, but with extraordinary attention – typical of the architect's work – to the construction details and a careful study of the proportions. Much more interesting was the second solution, which is undoubtedly one of the best of those presented in the competition. It proposed to articulate the bare main elevation with the insertion of nine trapeze-shaped bow-windows – clearly derived from the arrangement of the project for the Moscow MOGES – which guaranteed excellent lighting of the rooms. A similar solution was presented, outside of the competition, by N. Gundobin.
30. *Sovremennaja Architektura*, No. 6, 1929, pp. 186-210.
31. G. Orlov, "Dneproges vozrozhdaetsja," *Architektura i stroitel'stvo*, No. 21-22, 1946; G. Orlov, *Architektura vosstanavlivaemoj Dneproges, sbornik Architektura promyshlennick sooruzhenij*, izd. Akademija Architektury SSSR (Moscow, 1949); *Vseobshchaja Istorija Architektury*, t. 12, I kn. (Moscow, 1976), p. 353. The reconstruction project was developed by G.M. Orlov with V.A. Vesnin as a consultant. The new postwar buildings of the hydroelectric complex were designed by a collective coordinated by Orlov himself, and composed of P.L. Ryzik, T. Orlova, Ju.N. Gumburg, E.L. Beketova, A.Ja. Mosinskij.
32. V. Perlin, "Architektura verchnevolzhskich gidrouzlov," in *Architektura gidroelektrostancij* cit., pp. 95-140; *Vseobshchaja Istorija Architektury* cit., p. 348.
33. See for example E. Popov, "Problemy tipovogo proektirovanija promyshlennich predprijatij," *Architektura SSSR*, No. 10, 1953, pp. 3 ff.
34. For their architectural definition, one of the period's largest competitions was convened.
35. A. Kovalev, "Architektura sooruzhenija Cimljanskogo gidrouzla," *Architektura SSSR*, No. 2, 1951, pp. 2 ff.
36. *Voprosy architektury godroelektrostancii: Materialy nauchnogo soveshchanija* (Moscow, 1952).
37. See A. De Magistris, "URSS, l'altra ricostruzione," *Rassegna (La ricostruzione in Europa nel secondo dopoguerra)*, No. 54, 1993, pp. 76-83.
38. E. Popov, "Problemy razvitija architektury promyshlennych sooruzhenij," *Architektura SSSR*, No. 9, 1955, pp. 10-17.

Unless otherwise indicated, all the pictures are taken from the AEM Archives, Turin.

1. Construction site of the Ceresole Reale dam, 1928.

2. Watershed of the Ceresole Reale dam, at the beginning of the filling, 1930.

1

Patrizia Bonifazio

From the Valle dell'Orco to Turin: Production and Transportation of Electrical Energy

Problème technique simple : de la patience et de l'exactitude pour relever les niveaux de la vallée et des ses versants. Une multiplication pour cuber l'eau du lac artificiel qui serà créé. Un peu de règle à calcul pour résoudre quelques formules relativement simples. On conclut : il faut élever un barrage de tant de mètres de long, tant de mètres de haut ; il aura telle épaisseur à la base, telle au sommet, la poussée sur le barrage étant de tant. Un esprit moyen peut faire le tour de ces calculs: étape insignifiante.[1]

At the meeting of 5 February 1917, the exploitation of the Orco river was proposed.[2] On the agenda, the meeting had the government concession for the exploitation of the waters of the Lys to the Lombard company Breda and the consequent transportation of the energy obtainable outside the limits of the district of Turin. At the following meeting, a commission was set up to evaluate the predicted energy requirement for postwar industrial development – estimated at around 50,000 Kw[3] – the council decided to begin the paperwork to join the district council in the project for the exploitation of the Orco, the closest to Turin of the water courses available at the time (only 63 km from Rosone, in the lower valley, to the distribution station of Martinetto in Turin).

The problem of the production of electrical energy certainly did not affect the municipality because of its parochial implications: the lively speeches at the meeting of the council on the presumed offence to the city concerned the double role of the municipality, as administrator and producer of electrical energy.

The debate on the autonomy of local authorities which, with different intensities, characterized the last part of the 19th century all over Europe,[4] and which led in Italy in 1903 to the defining of the law on the municipalization of public services, led in Turin, as in Milan or in Rome, to the administration by the Municipality of the electrical company as well as the streetcars and buses. The exploitation of the concession on the Dora Riparia River between Salbertrand and Chiomonte made an initial output of 12,800 HP available; in 1911 this reached the production of 18 million kWh for 4673 users, enabling the company to set a tariff of 70 cents, compared to the initial 90.[5]

Yet certain observations are necessary. The presence on the municipal council of engineers and industrialists, or rather, of engineer-industrialists, from 1903 marked a real transformation of Turin into an "industrial city."

Placed at the center of speeches and debates was the problem of how to organize the territory to promote productive investments

2

and increase the productivity of industries: the request for an agreement with the Chamber of Commerce, "a place where big industrialists are represented, and can express their opinion on the subject," the creation of a commission "composed of technical, industrial and administrative elements,"[6] offer an initial reflection on this theme. The municipality was one of the places where decisions were taken on the development and the management of the city.

Debates on municipalization – which in Turin *La Riforma Sociale* of Luigi Einaudi has a rich display of cues and presences[7] – must also be related to a situation where the research by hydraulic engineers and the construction of services for industry constituted a point of contact between the municipal council, the school of application of engineering, researchers and entrepreneurs. An extended scientific community, which was to find its own ground for consolidation precisely in the debate on the introduction of electrical energy.[8]

Furthermore, the pressure on the municipalized companies, especially in the electrical sector, increased considerably from the 1920s on and the advent of fascism did nothing but make the situation sharper: in 1926 the publication of a magazine called *La Municipalizzazione in Italia* (published, not coincidentally, in Turin),[9] to which the by then numerous municipal companies producing and distributing electrical services subscribed,[10] underlines the tension between private groups and public administrations, between public administrations and corporate state, in a situation oscillating between autonomous management[11] and governmental management,[12] which was to drag on until at least the end of World War II.

Beginning at the turn of the century, the request for concessions for the exploitation of the Orco River was the object of numerous applications, from both private bodies and the Turin district government itself.[13] The possession of water to create energy, at least until the 1930s, marked a full-blown war over the securing of the rivers: to restrict ourselves to the companies on the Turin market, those present were Società Idroelettrica Piemonte (SIP), formerly Elettrochimica Pont-Saint-Martin (1918), Società Piemonte Centrale di Elettricità (1912), Società Idroelettrica del

3. Drain gallery of Ceresole Reale dam, 1929.

4. Pressure water pipe of the Rosone power station, 1929.

Monviso (1918), Elettricità Alta Italia (1896), Società delle Forze Idrauliche del Moncenisio-Sfim (1900), Società per le Forze Idrauliche dell'Alto Po (1902). From the 1920s until the concentration in SIP in the 1930s, Società Idroelettrica Piemontese-Lombarda Ernesto Breda (1920), Società Idroelettrica dell'Isarco, Elettricità Val di Susa, Idroelettrica di Stura (1925), Idroelettrica delle Marmore, Idroelettrica dell'Evançon, Gestione Officine Elettriche (1924) were founded. These producers and distributors, from the Lys Valley to Lake Moncenesio, from the Evançon to the Po River to the Tanaro River, fragmented the territory into a maze of systems.[14] A network designing a "geography" also related to economics, banking and industry: to remain in the Valle dell'Orco, in 1917 SFIM tried to take over the private concessions and Società Elettricità Alta Italia (EAI) made direct agreements with the Turin District Administration. Controlled by foreign capital, these two companies came within Fiat's orbit: in 1917 Fiat made agreements with EAI for the exploitation of the Orco River and at the same time with the municipal administration;[15] in 1919 SIP bought EAI; in 1919 Fiat bought a conspicuous package of SFIM shares, which were to be sold to SIP in exchange for participation in the company.[16]

The municipal council seemed forced to mediate, therefore, with a company such as Fiat first and with the Turin District Administration later, with which it was to begin a long dispute that was only to be concluded in 1921.[17]

The need to avoid risking new private interference, but also the possibility of maintaining a contractual weight,[18] at least as far as the supply of the service was concerned, therefore led to the presentation in April 1918 of the general project for the construction of the Valle dell'Orco hydroelectric plant.[19]

The plant envisaged the construction of large water reservoirs for seasonal accumulation to keep the supply of energy constant. The reservoirs were to use the derivations from lake Serrù and the Agnel River upstream from Ceresole Reale, that of the Orco with the outlet under the built area of Ceresole Reale and return at Rosone, that of the Piantonetto with outlet from the reservoirs of Pian Telessio and from the lakes Balma and Eugio, that of the Orco with outlet at Rosone and return at Bardonetto Inferiore. The plant was to cover a surface area of 250 sq km, exploiting an overall drop of 2780.76 m according to the derivations, and would produce a total of 32,000 kW referred to Turin, corresponding to an annual

3

4

production of approximately 300,000,000 kWh. The financing of the project – totalling 120 million lire – was to be carried out thanks to those same interests of capital provided by the municipalized companies in Turin – the municipal electrical company, the streetcar and bus company, the municipal aqueduct – which in 1919 had assets of around 18 million lire.[20] In the presentation of the project, the plant thus conceived was not only to allow the best use of the waters, but, added to the production of the plants of the Valle di Susa, it was to produce 72,000 HP in real terms, with an output corresponding to 50,000 kW and over 400,000,000 kWh produced in the thermal power generation plants.

Of the reservoirs to be realized, the first to be built – to allow the gradual functioning of the plant – and the largest, was to be that of Ceresole Reale, 3-km long and averaging 600-m wide, with an envisaged capacity of 14,000,000 cubic meters, from which the upper derivation of the Orco was to have begun. Its waters were to merge directly into a channel which, flowing along the bank of the Orco for 16 km, was to reach the loading tanks at the Piantonetto and, through high-pressure water piping and a transfer of 813.55 m, the Rosone power plant, producing an output of 32,542 continuous rated HP. The project for the dam was presented in 1924: the dam which was to be built was a gravity one, with a guard wall, as was suggested by the Commission for dam construction. The type with a triangular section and the upper tip at the level of maximum storage, protected by the inclined arches of the guard wall, was the one which offered the best guarantees of safety, the possibility of best counterbalancing the force of the water, but above all the possibility of continued supervision. A decision that perhaps should not be evaluated only on its technical aspects: in the construction of the dam system, the requirement for safety was the only element that the municipalities of the area – such as Noasca, below the reservoir – called for, under penalty of expropriation of the land for public use.[21]

The dam and the storage lake, which were superposed over an existing land system, constituted the visible element of a complex system, formed also by a non-visible part that changed the layout of the mountain territory.[22] A change that is suggested if we view the chorographic map of the location. Visibly, it is the construction of the site, even before the transportation of the energy to the city, that transformed the landscape.

5

5. Ceresole Reale dam, 1929.

6. Chorographic map of the Orco Valley with the project of the plant. From: C. Bornati, *Gli Impianti Idroelettrici del Municipio di Torino nell'Alta Valle dell'Orco* (Turin, 1930).

In 1921 the work began on setting the infrastructures in the land, the construction of those secondary works, which were to allow the transportation of the materials for the construction of the high-altitude dam structures and the self-sufficient management of the site. The construction of the access roads from Pont Canavese to Ceresole saw the intervention of the Municipality in collaboration with the municipalities of the valley. Technical reasons led to the redefining by the Municipal Hydroelectric Plant Service of whole sections of road, such as the deviation of the Sparone development, but they also made the valley the object of studies for a reduced gauge railway and a trolleybus line linking Pont to Prà-Lilla. To supply the site itself, a provisional power plant was built in Noasca, with an output of 1000 kW. A telephone line connecting the six municipalities of the valley was installed.

A transformation that is marked by the change of use of the land, which through contracting proprietor by proprietor allowed the construction of small reservoirs in the valley, deviations in the course of the Orco for the irrigation of the fields and animal breeding.

In 1925 the construction of the dam was begun. On 1 May 1929 the Rosone power plant was inaugurated. The electroducts connecting the Rosone power plant and through the Canavese region to reach Turin began operating in August 1929. On 2 August 1931 the Prince of Piedmont inaugurated the dam.

The return of electricity to the city strengthened the distribution network of the conversion plants: in 1929 the Sebastopoli sub-station[23] was built, to the south of the city, in what after the building of the Lingotto car plant in 1920 and the inauguration in 1939 of Fiat Mirafiori was to all intents and purposes to take on the identity of a large "grey area," capable of redefining globally the functional and physical organization of the city. In 1932 the Vanchiglia and Aurora sub-stations to the north were ready. The electrical network did not make provision for certain localizations and consumptions: it followed production requirements that were changing – consider the innovation of the continuous cycle at the Lingotto and the modifications that this brought to the land through the energy required and human and technical resources moved[24] –, responds to the requirements of collective life, linked increasingly to production times.

The electrical production of Rosone, already from 1929 met with a not irrelevant number of problems of discharge: competition with SIP, for the control of its users, not only caused a charging situation that was anomalous in the whole of Italy, but involved almost a house-by-house war, due to the situation, which was curious, to say the least, of managing two overlapping distribution networks. In 1931 this situation found a solution with the agreement between Azienda Elettrica Municipale (AEM) and SIP, not only on the sharing of users and the application of fees, but also on the purchasing by SIP of quotas of the energy surplus produced by AEM.[25]

Far removed from the autarchic rhetoric, the solution found to absorb the excess was that of domestic consumption, for heating above all, with equipment using steam and hot water. An exhibition and sales room, climbing aboard the Futuristic image, in Via Roma Nuova – the emblem of modernization –, advertising on the streetcars and buses, in the newspapers, on matchboxes, advertising stands, practical experiments at the annual Metal Work Exhibition, the granting of premium vouchers to advertising users, all promoted the use of electrical energy. Normal resistance and Thovez heaters, hotplates, floors, walls, radiant ceilings, air heaters, storage heaters heated municipal schools and hospitals, the Turin Museum of Ancient Art in Palazzo Madama. Rented water heaters were installed in 1200 homes. A standard cooker was created for "the lower classes," with two rings, "also payable in twelve monthly installments, including, along with the hob, the socket, internal installation and a series of flat-bottom saucepans,"[26] all so that the light might be.

6

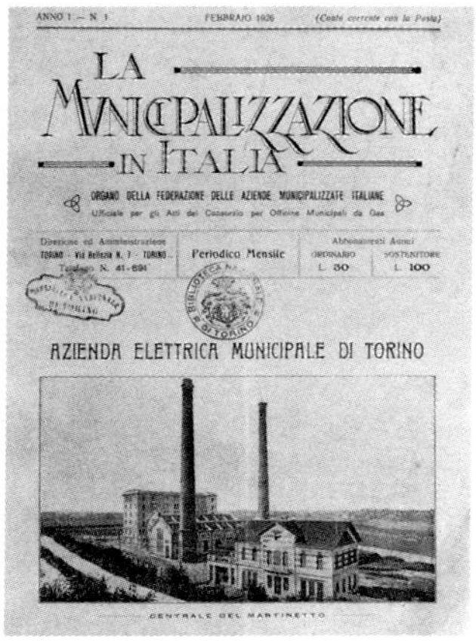

7

7. *La Municipalizzazione in Italia*, No. 1, February 1926. Front cover.

8

8. Martinetto power station in Turin, 1930.

9

9. Martinetto power station: inside view.

1. Le Corbusier, *Urbanisme* (Paris, 1925), p. 138.
2. Historical Archive of the City of Turin (ASCT), *Atti del Municipio 1917*, minutes of the meeting of 5 February 1917.
3. *Ibid.*, minutes of the meeting of 4 June 1917 and the meeting of 22 August 1917.
4. In the extensive bibliography on the theme, see D. Calabi, "I servizi tecnici a rete e la questione della municipalizzazione nelle città italiane (1880-1910)," in P. Morachiello and G. Teyssot (eds.), *Le macchine imperfette. Architettura, programmi, istituzioni nel XIX secolo* (Rome, 1980), pp. 293-316; R. Franco, "Il dibattito sui pubblici servizi e le municipalizzazioni alla fine del sec. XIX," in *Storia urbana*, VI, No. 20, 1982, pp. 77-93; P. Dogliani, *Un laboratorio di socialismo municipale. La Francia (1890-1920)* (Milan, 1992).
5. R. Balzani and A. Varni, "Le aziende elettriche municipalizzate," in G. Mori (ed.), *Storia dell'industria elettrica in Italia. Le origini, 1882-1914*, vol. 1 (Bari, 1992), p. 559. Cf. also P. Hartner, "L'esperienza della prima impresa pubblica nel settore elettrico: le aziende municipalizzate in Italia, il caso tedesco e altri modelli stranieri," in *La nazionalizzazione dell'energia elettrica. L'esperienza italiana e di altri paesi europei. Atti del convegno internazionale di studi del 9-10 novembre 1988 per il XXV anniversario dell'istituzione dell'ENEL* (Rome-Bari, 1989), pp. 73-89.
6. The commission was composed, among others, of Teofilo Rossi, mayor; municipal councillors Giuseppe Pomba, Secondo Frola; of Ferdinando Bocca, Chairman of the Chamber of Commerce; Dante Ferraris, Chairman of the Industrial League. Cf. ASCT, *Atti del Municipio 1917*, minutes of the meeting of 4 June 1917.
7. The presence of the magazine in Turin does not lead, however, to its taking an interest in local matters. Turin appears as the subject, along with other Italian and foreign cities, of the investigations that the magazine was conducting at the time.
8. C. Olmo, "Torino da città capitale a città industriale," in *Studi Storici*, XXIV, No. 1-2, January-June 1983, pp. 217-35.
9. The actual editor-in-chief of the magazine was Giuseppe Bisazza, engineer, director of the AEM in Turin; the deputy chairman was Cirillo Rossi, engineer, director of the municipal services company of Vercelli. The magazine changed in 1927 into *Le Industrie Municipalizzate. Bollettino della federazione fascista delle aziende industriali municipalizzate*.
10. In the "Indagini statistiche sulla produzione dell'energia elettrica in Italia" in March 1928 there appeared, as well as Turin, Milan and Rome, in order of size, Fiume, Merano, Naples, Verona, Bressanone, Brescia, Tolentino and Parma: *Le Industrie Municipalizzate*, II, No. 4, 1 April 1928, p. 84.
11. Cf. for example Giarratana, "Esagerazioni o peggio contro le aziende municipalizzate," in *Le Industrie Municipalizzate*, II, No. 10, 1 October

10

10. Showroom for the AEM electricity domestic applications, Turin.
From: *Le Industrie Municipalizzate*, No. 3-4, April-March 1935.

11. Turin map indicating the stations of the electric distribution, 1932.
From: *Le Industrie Municipalizzate*, No. 1, January 1933.

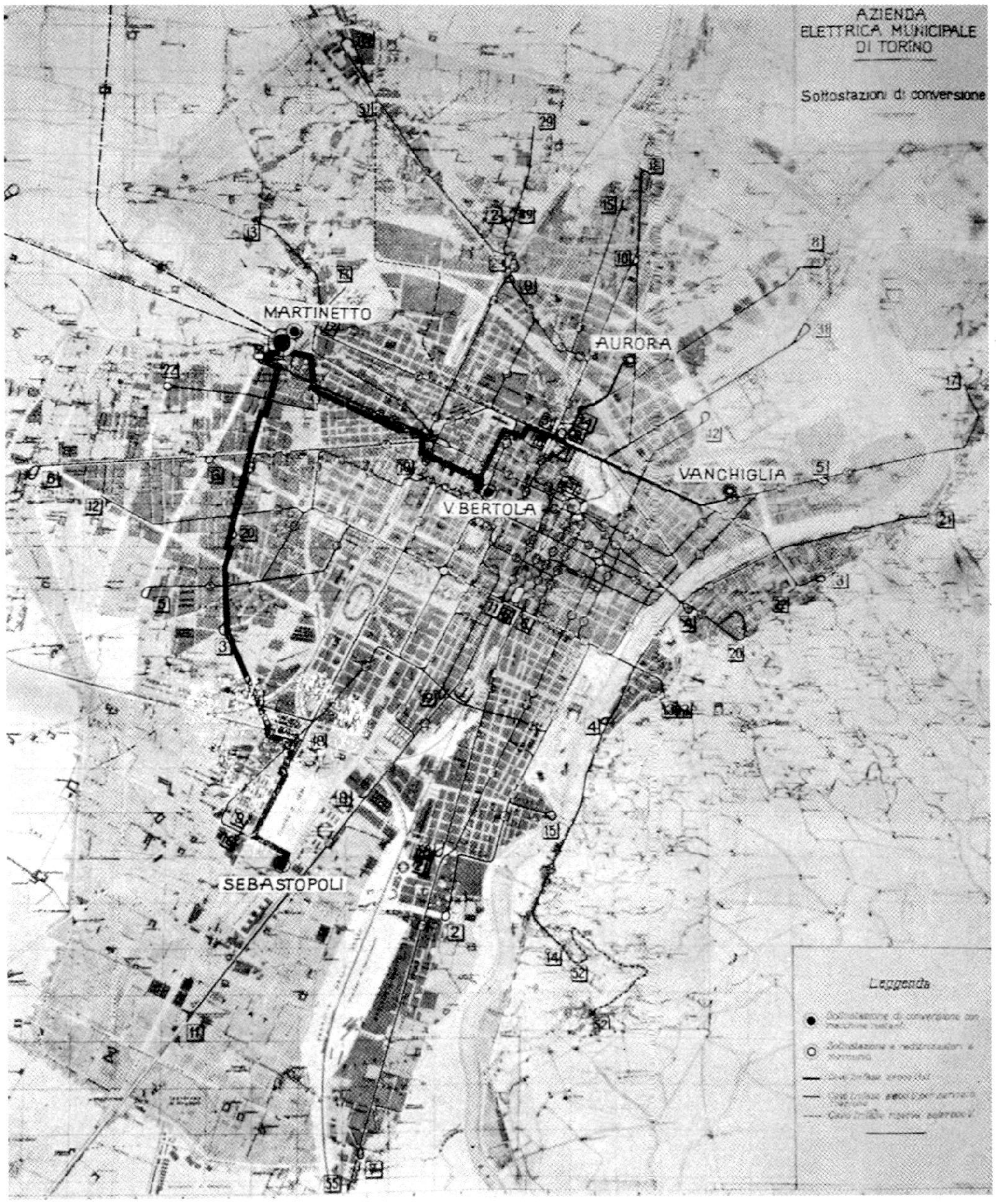

11

1928, pp. 94-5; G. Monleone, *Il movimento municipalizzatore e le aziende municipalizzate durante la guerra e nel dopoguerra* (Rome, 1925) and the interesting Id., *Bibliografia generale delle municipalizzazioni* (Milan, 1927).
12. Cf. "Per un migliore assetto della produzione dell'energia elettrica in Italia. (L'opinione dell'ing. Giacomo Levi)," in *Le Industrie Municipalizzate*, II, No. 4, 1 April 1928, pp. 78-80.
13. Requesting the concession were Count San Martino d'Aglié, with a project that envisaged the exploitation of the derivation between Ceresole Reale and Rosone; engineer Barberis, for the derivation of the Piantonetto, Lake Balma and Lake Eugio; Martinazzi for the formation of an artificial lake downstream from Ceresole Reale. More complex, due to the size of the territory involved, was the project presented by the Provincial Administration, formulated by its engineers Corazza and Soldati, which set itself the task of exploiting the whole hydrographic basin of the river and proceeding with reforesting it: cf. C. Bornati, *Gli Impianti Idroelettrici del Municipio di Torino nell'Alta Valle dell'Orco* (Turin, 1930), p. 1.
14. The data is taken from C. Pavese and P.A. Toninelli, "Anagrafe delle società elettriche: la documentazione di base," in G. Galasso (ed.), *Storia dell'industria elettrica in Italia. Espansione e oligopolio. 1926-1945*, vol. 3, t. II (Rome-Bari, 1993), pp. 1066-258.
15. ASCT, *Servizio tecnologico industriale, Derivazione delle acque pubbliche, Derivazione dall'Orco, Pratica colla Soc. "Fiat"*, Class 93, Pos. 4, File 4, 1918.
16. A. Castagnoli, "Il passaggio della SIP all'IRI," in G. Galasso (ed.), *Storia dell'industria elettrica in Italia* cit., pp. 595-642.
17. A dispute that concerned the payment of a quota for the expenses sustained for the studies on the valley and the transfering of a part of the energy obtained from the exploitation of the river: C. Bornati, *Gli Impianti Idroelettrici del Municipio di Torino* cit., pp. 6-19.
18. Faced also with the autonomy of the companies in the choice and production location: cf. B. Bianco and A. Frisa, "L'operazione Ansaldo-Silp a Torino: una grossa proprietà fondiaria di origine industriale nel processo di costruzione della città," in Città di Torino, *Torino tra le due guerre* (Turin, 1978), pp. 100-21.
19. All data concerning the phases of the site and the technical data on the construction of the plant are taken from G. Bornati, *Gli Impianti Idroelettrici del Municipio di Torino* cit.; "Gli Impianti Idroelettrici del Municipio di Torino nell'Alta Valle dell'Orco," in *Le Industrie Municipalizzate*, V, No. 4, 1931, pp. 128-43; "Gli Impianti idroelettrici del Municipio di Torino nell'Alta Valle dell'Orco," in *L'Energia Elettrica*, III, No. 5, 1931, pp. 385-98; No. 6, 1931, pp. 465-77.
20. Actually, it was to be the Cassa di Risparmio that supplied the necessary funds, in exchange for bonds of the Consorzio di Credito per le Opere Pubbliche in Rome.
21. ASCT, *Derivazione dall'Orco, Opposizione dei Comuni della zona*, Class 93, Pos. 4, File 4, 1921.
22. Cf. D. Geissbühler, "Barrages dans l'espace alpin," in *Faces*, No. 22, Winter 1991, pp. 21-31.
23. M. Brunetti, "Nuove stazioni di conversione dell'Azienda Elettrica Municipale di Torino," in *Le Industrie Municipalizzate*, VII, No. 1, January 1933, pp. 1-14.
24. Cf. C. Olmo (ed.), *Il Lingotto, 1915-1939. L'architettura, l'immagine, il lavoro* (Turin, 1994), pp. 9-42.
25. A. Castagnoli, "Il passaggio dalla Sip all'Iri" cit., pp. 611-3; "L'Azienda Elettrica Municipale di Torino e la convenzione con la Sip," in *Le Industrie Municipalizzate*, II, No. 10, 1928, pp. 208-14.
26. M. Brunetti, "Riscaldamento elettrico e applicazioni domestiche in relazione a eccedenze di produzione idroelettrica," in *Le Industrie Municipalizzate*, IX, No. 3-4, March-April 1935, pp. 41-53.

Anna Maria Zorgno

Electricity and Land

In the landscape understood as a physical, built form of land, nature and artificial elements alternate, giving rise to numerous possibilities of "descriptive" and perceptive experiences. Before the idea of human progress is replaced by that of technological progress, the result of man's interventions in the land – buildings, road and canal systems, but also materials, spaces and colors – express the more mature stages of a certain kind of culture, rooted in the physical environment and associated with the resources of that ecosystem, a part and an expression of a rich and balanced ecosystem, stopped in one of the phases of its transformation. Other possible interpretations privilege, in the face of the "description" of a built landscape (or some element or characteristic part of it), the perceptive aspect: the observed object then assumes a value solely as an image, remaining foreign to any practical purpose or research of content: "a part of the land, including the sky above it, becomes a landscape when we accept or discover it as a sensible experience that has actually been experienced, which is conveyed to us as an impression."[1]

In a land which has been marked by the itineraries of the great utilitarian routes – fortifications, shipways, aqueducts, roads, railways – the novelty of a design which, until the late 19th century, was understood as a proposed integration between technical intervention and nature, is associated with new possibilities of representation of the space, made possible by new technical means of observation and new points of view which some of these same routes contribute in creating. The resulting possible perceptive approaches are characterized by an emphasis of the unitarian effect, and the quick succession of a large number of images gives an illusion of a greater depth in the representation of the designed space.[2]

The description and perception of the built landscape refer to the major processes of infrastructural reorganization of the land, elements which can be identified directly: routes, or itineraries, which are tangible and concrete, and which model the landscape, build it, mark it in relation to the requirements and needs of man, as well as to his culture, and which reflect – on the different occasions – new possible expressions, or definitions, of natural qualities. Until the middle of the last century the construction of the landscape was characterized by a human dimension; then, with the creation of new forms of mobility, of production, of urban culture, and above all, with the prevalence of the strategy of flows and speed, the anthropic landscape began a process of gradual deterioration, while artificial elements and nature ventured on different and often antithetic paths.

A relationship vacillating between encounter and conflict was thus created between nature and technical artificiality. This relationship has led to the creation, within the landscape, of a new historical condition for the relations between the individual and progress, between society and land development. The great narration of progress then resorts, so as to legitimate itself, to new parameters of critical evaluation, such as that of "artificial quality," whose definition is completely associated with an instrumental concept of the physical space.[3]

The common fate of the elements produced in lands marked by these routes, both in the anthropized landscapes and in those crossed by infrastructural networks since the middle of the 19th century and later, is that of a slow transformation over time; but also that of presenting paths of research aimed at a possible identification of traces, fragments, of subsequent sedimentations. This is an extraordinary field of study for the architect, due to

1

1. Gabriele Basilico, Po Lowland, 1989.

2. Executive drawing of the supporting metal tower consisting of riveted iron section bars (Società Nazionale delle Officine di Savigliano, 1928). Archives Azienda Energetica Municipale, Turin.

the richness of the meanings and the suggested references which these traces propose on the different occasions. A description and interpretation of these traces may in fact facilitate a recomposition of those particular networks, including those which do not belong to the natural context, with which the landscape has expressed its identity, its involvement in mixed, but determining uses, chosen by man, as a result of his being part of a particular civilization.

On the built land the typical structure of the electrical systems of production, transmission and distribution of energy has proven to produce, from the beginning, different levels of incidence. The power lines have revealed, since the years when the problem of transmitting electrical energy over distances was first posed, a characteristic resistance to integrating with the built land, to suggest possible readings – or interpretations – according to the new orders, or the different strategies associated with models of artificial quality that followed the change.

Since the technical and scientific experiences made in the late 19th century, the most promising ones for the solution of the problem of the transmission of electrical energy over distances (since the rapid progress made in this field during the early years of our century is attributable to these developments), the greatest concern of designers and researchers has been to identify methods that made it possible to optimize the technology of electrical transmission, according to a research that was to produce the most mature results in the years after World War I. The engineering of electric transmission systems, which precisely in the early 20th century became an autonomous "design" discipline, tended to radically isolate the technical issues from those of land interference. The problem of designing power lines was entirely centered on the new discipline, electrotechnics, and the importance of the issues that had to be resolved with the help of the specialized competences it contributed to creating were such as to overshadow other problems. The land, crossed by power lines which covered increasing distances, represented a passive material awaiting the installation of new lines, while the debate focused solely on technical issues; the dominating theme – the crossing of the greatest distances possible at the greatest profit – was associated with the debate as to which was preferable, direct current or alternating current.[4]

Nor could it be otherwise, if we consider the long period during which the study and re-

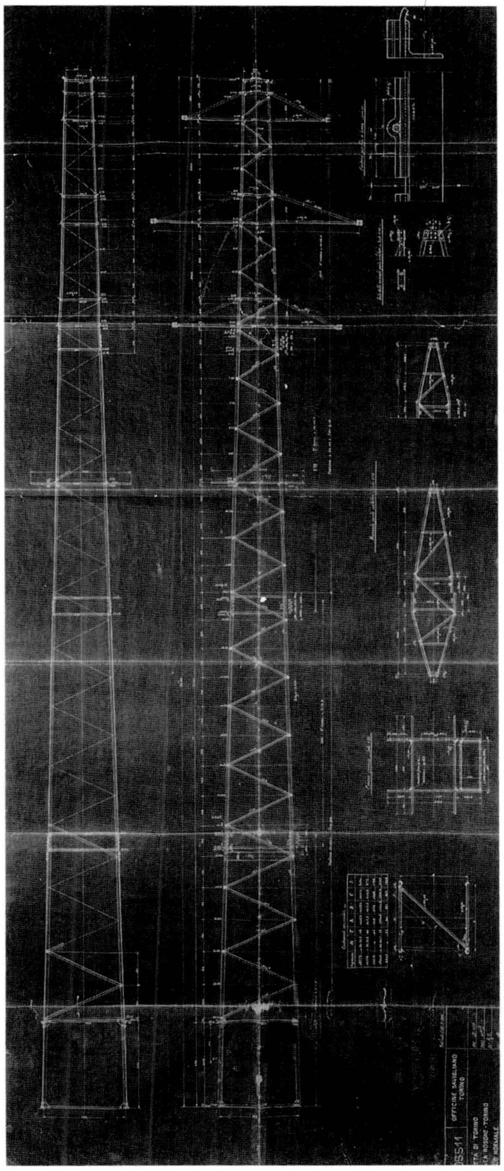

2

search on the transmission of electrical energy over distances had been carried out in Europe – from the 1870s up to the 1930s – almost always venturing on daring and far-sighted initiatives, which were nevertheless of an almost exclusively financial nature, and especially if we turn our minds to the important function of guidance which, already since the middle of the last century, the electrotechnics sector exercised on the different fundamental aspects of life. This guidance function had been manifest not only in the transportation of electrical energy, but also in the realization of direct or alternating current generators and in the construction of electrical engines.[5,6] All these problems called for a systematic use of energy on the part of technicians, but especially on the part of physicists; and in fact electrotechnology was to succeed in solving them under the guidance of the most modern scientific research in the field of physics.[7]

Precisely with regard to the subject of transmission of electrical energy over distances suffice it to consider the scarce interest devoted to alternating current at first; one did not know how to use it, and in particular how to vary its tension, nor did one know how to transform it into mechanical energy, at least not until the mid-1880s.

Only as of those years, thanks to the realization of the first engine with rotating magnetic field created by Galileo Ferraris, could the methods of transmission of electric energy over distances greatly benefit from the possibility of stepping up or down the voltage of an alternating current made possible by "transformers" (on the occasion of the International Exposition of Turin in 1884 Lucien Gaulard demonstrated a transmission of alternating-current energy by powering, with the machines installed on the exhibition premises, a number of electric bulbs located at the Lanzo railway station, 34 km away).[8] It was now possible to take advantage of a very simple principle to transport energy: high-voltage currents could in fact be transported over long distances with less energy dispersion; at destination their tension had to be stepped down considerably in order to make them usable without inconveniences.

In the early 1880s electricity therefore left the laboratories, shedding the trait of scientific curiosity which had characterized the first stages of research; the first modern cathedrals were built for its production as well as the first stations for its transportation, also thanks to a know-how which had by then been fully mastered by applied research, through the development of instruments which served for the maximum exploitation of the new energy carrier. The beginning of the supply of electrical energy was in fact associated with commercial exploitation and the powering of urban services. In Italy electrical energy for commercial purposes became available with the inauguration, in late June 1883, of what the national literature defined as "the first European power station," the Edison station of via Santa Radegonda in Milan, next to the Cathedral, shortly after the Pearl Street station in New York, planned and built by the Edison compa-

3

3. Supports in the laguna. Società Italiana per l'Utilizzazione delle Forze Idrauliche del Veneto, 3000 V main line (1905).
From: Associazione fra Esercenti Imprese Elettriche in Italia, *Notizie sui principali impianti elettrici d'Italia* (Milan, 1911).

4. Supports for the Paderno-Milan line (1898). Società Italiana per Imprese Elettriche.
From: *ibid.*

5. Supporting towers of the Società Elettrica della Sicilia Orientale, Siracusa-Catania-Messina line, during the assembly stage (1910).
From: *ibid.*

ny itself the year before, and the inauguration, also in 1882, of the Edison plant in London, the Holborn Viaduct.[9,10]

But only during the last decade of the century did a new phase in the life of the electrical industry, characterized by the new possibilities opened by the method of transportation over distances, take place. The favorable prospects made feasible in this sector by technical and scientific progress became evident at the Exposition of Frankfurt in 1891. The age of transportation over distances was also the age of an increasing use of electrical energy as motive power, as well as a light source, and the age of the building of dams and power lines which made an increasingly dense net over plains and mountains, giving the land a new, sometimes unsettling, technical artificiality.[11]

At the end of the century the transportation of energy ceased being an exclusively technical problem, as it had proven to be during the previous decade, in order to become a true manufacturing initiative. The most important among the earliest hydroelectrical plants built in Italy at the time became international sensations, and have remained a legend among Italian electrotechnicians: the Tivoli-Rome plant and the Paderno-Milan plant, first-class enterprises from a technical point of view, by virtue of their innovative conception, the complexity of the difficulties, the power transmitted and the distance covered by the line.[12,13]

The line from Tivoli to Rome, which covered 28 km and transported three-phase alternating current, was built one year after the realization of the Lauffen-Frankfurt line (1891), which was 175 km long and which also transported three-phase alternating current. The line from Paderno to Milan, started by the Edison Company of Milan in 1898, transported three-phase current at a tension of 14,000 V over 35 km. The latter line, especially, which for some time was to remain the most important energy transportation in Italy and abroad, posed serious construction problems. Due to the distance and the high tension completely new problems had to be solved in order to achieve mechanical stability, insulation and protection from atmospheric discharges of the lines and the supports, a true calamity for the first power lines. To guarantee sufficient stability it was necessary to use, for the first time in Europe, iron towers designed by Giulio Röthlisberger of the Società Nazionale delle Officine of Savigliano. American-type insulators as well as insulators studied specifically for the line provided the necessary insulation and protection.

4

5

In the first years of the 20th century the technology of transmission of electricity could be considered as clearly delineated and to a considerable extent mature with regard to all the main design problems. In those years one passed, in Europe, from a maximum tension of 20 kV to 80 kV on the eve of World War I, while 100 kV were reached as early as 1909 in the United States with the 250-km long line of Bing Bend-Oakland in California, and the 340-km long line between the Great Falls and Durhan in North Carolina built shortly afterwards.

After World War I tensions above 100 kV were rapidly developed also in Europe. In Italy the decade which preceded the World War was characterized by great expansions in terms of production and constant amplification and in-depth research in the field of electrotechnology. In addition to the design and construction of increasingly powerful plants, located at ever greater distances from the centers of use, in those years Italian technicians carried on studies and research on an increasingly wide range of subjects, extending their theoretical knowledge. Significantly, the most studied subject in that decade, due to a growing demand for energy, was the power lines. The improvements achieved with regard to design calculations and organization of the builder's yards led to very advanced constructions, such as the Grosotto-Milan line connecting the AEM stations in Valtellina, a line which, in addition to the considerable length (150 km) included stretches of high mountains, which posed completely new problems, especially due to the load of snow and ice; one began to tackle the problems of mechanical stability, of power leakings and tension dropping, in order to achieve economical optimization.[14] While this objective was never refuted since then, it was not associated with an awareness of the substantial transformations undergone by the land crossed by power lines until much later.

The attention devoted to transmission lines in those years is explained by their high cost which, as the distances became greater, had become a very significant part of the investment necessary for the construction of a plant; likewise, the problem of protecting the lines from tension overloads of internal and external origin and discharges to the ground had become the principal problem for the continuity of the service.

In Italy the first 130-kV line was built by the Società Interregionale Cisalpina in 1923: it connected Borgherio (Milan) and Reggio Emilia, covering a distance of 153 km. The year

6. Metal tower during the assembly stage during the first decade of this century.
From: L. de Rosa (ed.), *Storia dell'industria elettrica in Italia*, vol. 2 (Bari, 1993).

7. Special support produced by the Società Nazionale delle Officine di Savigliano in the approach to Milan on the Paderno-Milan line (1898).
From: G. Mori (ed.), *Storia dell'industria elettrica in Italia*, vol. 1, part I (Bari, 1992).

before 220-kV tension had been introduced for the first time with the Petit River (California) line of the Pacific Gas & Electric Company. In the years from 1929 to 1930, with the Italian line of Isarco (Società Elettrica Piemontese) and the German line of the Rheinisch-Westphälisches Elektrizitäts-Werk, 220-kV tension was introduced in Europe.

Parallelly, precisely during the 1920s, the organization of the distribution system through the building of power stations and the development of projects of urban and non-urban illumination led – directly or indirectly – to an artificial exploitation of the land, in order to create electrical networks, something which caused significant changes in its configuration. The large plants of the stations often entailed the deviation of river courses and the filling of loading basins through the creation of artificial canals; entire valleys were submerged by the water checked by the dams, the towers and pylons of the power lines enclosed the landscape in an increasingly dense grid, to which the original characteristics of the land opposed an ever weaker resistance.

The results of the research in the field of transmission of electric energy over distances was to produce the present, diffused high tension network, which connects most production stations, and which supplies an even more complex network of primary distribution, of reduced sizes, whose junctions lead to the secondary distribution network, and then to the low tension distribution system. Moreover, the development strategy was to start processes of conveyance to areas characterized by a high consumption of the electrical energy available in low-consumption, or even deserted, regions. The systems were then to show a fast evolution over time, proportionately to the tension, which varied with relatively fast dynamics, to adapt to a consumption that was to double every ten years, and even in less time in developing countries.

The network of lines, in this process of development and adaptation, was for a long time to remain indifferent to the transformations, not to say upheavals, caused to the land. Based on design strategies of a mainly technical character, it was to adapt its itineraries to policies of optimization of plant costs and performance, policies which had already been adopted by 19th-century railway engineers (with analogous results) who, when designing new railway lines, had traced the itineraries with silk thread on topographic maps, transferring the routes on the land as if it were an abstract support, only concerned with

6

7

obtaining as straight itineraries as possible.[15] In the relationship with the built land, beyond motivations of a technical kind, the metaphor of the immateriality of the air line has probably prevailed, along with that of the immateriality of the transportation of a particular kind of energy; a clean energy, precisely because of the fact that it is intangible, inconsistent, without dross.

And yet it is precisely these presumed characters of immateriality on the part of transmission lines, along with the awareness that the basic raw materials, to a large extent represented by water, possesses a characteristic docility, thus allowing itself to be molded, its course to be regimented, to be quantified in the basins, to the point that it, along with the energy it produces, becomes a design material, which have inspired design proposals aimed at involving very large parts of the land, although such projects belong to the context of the technical and architectural Utopia characteristic of the early 20th century. Among these, for example, "the greatest project of all times," the project of Atlantropa, prepared by German architect Herman Sörgel between 1927 and 1929 with the collaboration of a large group of technicians and engineers.[16] The project consisted of a complex plan of reorganization of the entire Mediterranean region by lowering the level of the Mediterranean sea in order to conquer farmable land and at the same time obtain important energy sources. It was based on the barrage of the Mediterranean by means of a large dam that would cover the 14 km from Tangiers to Gibraltar, which would have supplied energy to the world's largest hydroelectric station. Other power stations, however, were planned on the land that would surface by lowering the level of the entire Mediterranean sea, on the Dardanelles strait, in correspondence with the estuary of the great rivers, between Sicily and Africa (with the construction of the dam of Tunis). In a few years Sörgel's Utopian proposal was supported by important advocates. Among those who declared themselves to be his supporters we can mention some highly qualified German institutions, public and private, such as the Ministry of Labor of the Weimar Republic, the entrepreneurs Krupp and Siemens and the Bund Deutscher Arkitekten.

But Sörgel's project was not the only one in those years that saw the involvement of large parts of the Earth in strategies of energetic exploitation. In the early 1930s the Atlantropa project belonged to a context of great hydroelectric projects, in America and in Europe,

8

8. Supporting towers of the unified type on a mountain route. Archives Azienda Energetica Municipale, Turin.

9. Supporting towers near the transforming station. Archives Azienda Energetica Municipale, Turin.

conceived to meet different but interdependent objectives: production of electrical energy, protection of the coasts, the freeing of new land by subtracting it from the sea. In this period electricity and land gave rise to a new generation of technical utopias, symbolic image of the acme of the second industrial revolution: a revolution characterized, as well as by the progress of the chemical industry, by the mass electrification of the industrial countries, along with a parallel increase of the mobile capital.

Among the "realistic Utopias" whose ambitions were not far removed from those of the Atlantropa project we find the project for the Ijsselmeer in Holland – whose validity in terms of technical feasibility was to be demonstrated in 1932 when the dam was built – or the project of the Tennessee Valley Authority, inspired by Roosevelt's New Deal, but also numerous great projects launched in the 1920s to exploit hydraulic energy in the Soviet Union. But while almost all these projects were the result of the by then inevitable need to extend the electrification to the industrialized countries, Sörgel's proposal was, in comparison, characterized by specific political values: in fact, the Atlantropa project prefigured the construction of a true federation of European States, whose political and economic power depended on the exploitation of electrical energy. In the world of the future, "without coal," imagined by Sörgel, the countries participating in the Atlantropa project would possess, thanks to the power stations built along the new perimeter of the Mediterranean, the largest and most inexhaustible energy reserves in the whole world.

In this project nature and technology, territory and built landscapes are not treated as opposites; on the contrary, technology is considered as a second nature, and nature – while subordinate to artificiality – is welcomed as a source of energy. The scientific and technical revolution which took place in that period has not only changed both nature and its image, but also the perception of this image, as for example the one conveyed by the decorative arts, which adopted parallel processes of abstraction. In his mural *Le transport des forces* at the Palais de La Découverte at the International Exposition of Arts and Techniques in Modern Life held in Paris, Fernand Léger in 1937 depicted a landscape powered by electricity, a "built" landscape which shows evident signs of the transformation caused by the permeating force of electricity, in particular stressing the characters of abstraction. The

9

variegated multiplicity of the assemblies borrowed from an industrial technology which was, to an increasing extent, aimed at exploiting electricity translated the multiple perspectives which the built landscape is now subject to, to the point of suppressing the very canonical rules of perspective and the scansion of the proportions it traditionally suggested. At the same time the natural landscape and the elements that traditionally represented its symbolic expression – earth, fire, air, water – conveyed the transfigured image given them as a result of their being studied by technology in order to sustain the transportation of forces, an image through which the landscape seemed to be dissolved in an abstract determinism, all the more unsettling, as more explicit an unexpected creativity might have appeared.[17]

But the transportation of electrical energy left concrete, often indelible, marks on the land, with which it was hard to establish a direct relationship, to construct an enchantment which would allow us to understand them as a liveable part of the built landscape, and to accept its new artificial dimension. The solid supports (the poles), the supports with a reticular structure (the pylons), the towers for the crossing of medium or long distances remained, as they have been conceived, as objects placed boldly on the land, witnessing an itinerary of forces with which we could only build a direct experience in abstract terms. And yet their scansion, the networks woven by the overhead wires they supported, created perspectives which were anything but immaterial, which permeated the landscape with the characters of the technical abstraction which generated them, altering its natural references, proposing alternative perceptions that were associated solely with the artificial. New metaphors of the "industrial emergence," the supports for power lines, in spite of the various attempts to dematerialize them, made especially after the explosion of the ecological problem, have revealed their inability to propose those extraordinary analogies which other emerging industrial factors have been able to benefit from. Smokestacks, for example, one of the first symbols of a production connected to the exploitation of energy, have succeeded in borrowing, from the typical characters of a built – and sedimented – townscape and landscape (like trelliswork, but also belltowers and minarets), singularly appropriate characters, not as a result of attempts to mask or dematerialize, but due to a natural continuity with the surrounding environment and the way its fabric was traditionally formed. But while a smokestack indicates the presence of an industrial concentration, to some extent becoming its symbolical image, the supports for power lines denote a diffused emerging reality (technical rather than industrial), made of standard, repetitive elements, which segment the route of one of the many branches of a network whose origin and destination we do not know, and perhaps are not required to know. The support, the sequence of two or three supports, which is all we can perceive in a normal perspective view, indicates a station (or perhaps an intermediate threshold) of a route which, more often than not, remains unknown. We can only imagine the stages of departure and arrival of this route (from the station to the city outskirts) but, beyond these feeble, hypothetical references, and the awareness that it indicates the existence of an equally immaterial strip traced by the route, it does not trigger our imagination, or enable us to participate in the new built landscape.[18]

However, if we consider that in countries characterized by a high density of electricity consumption the surface of the ground devoted to the passage of overhead wires is around 1%, we realize the entity of the problems of interference that overhead wires represent for the land.[19] These problems have, as a lower limit, those of the space occupied by the individual supports and the limits of height imposed by the building regulations, and as an upper limit the height of the aerial wires and their supports.

At the beginning of the period between the two wars the classical structure used to support power lines consisted of homogeneous riveted iron section, which was practically the

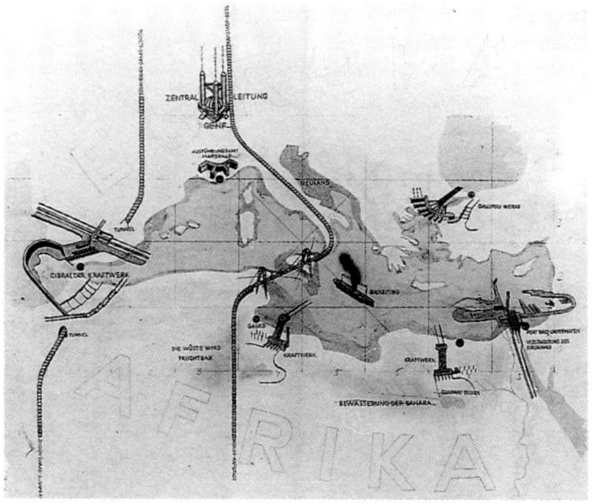

10

10. The Atlantropa project (1927-9). From: W. Voigt, "Le plus grandiose projet des temps modernes," in J.-L. Cohen and H. Damisch (eds.), *Américanisme et modernité* (Paris, 1993).

11. Supporting towers and poles. Banca Immagini ENEL.

only known type until 1925. A type which was rapidly improved, first with the use of high elasticity steel, then with the introduction of the tube, especially for the lateral poles, and finally with the adoption of welding for a considerable part of the connections. Until 1930, when this type of support, referred to as "very light pole," was adopted, 60 kV lines weighed about 15 t/km, and required 80-100 mc/km of concrete foundations; in the mid-1960s lines of the same weight, 15 t/km was used for 380 kV lines, including the metal foundations, i.e. without concrete. This significant reduction of the weight has become possible in those years by the increasing importance assumed by the electronic data processor in the design of supports for power lines, thanks to which the optimization of weight and costs has become a routine, by the systematic resorting to confirmations of the theoretical results by means of loading tests on the structures and by the wider template range of sections. An implicit priority objective underlies these studies: the research on the lightest, but above all the most economical pole, which embodies, in its typological and technological characteristics, the most favorable combination of all economic parameters: the cost of the materials, the cost of labor and the costs of transportation and assembly. This objective has never been disproved in later years; on the contrary, it has suggested the adoption of even lighter types of supports, such as the stayed pylon, with a few rigid elements hinged to one another and kept in place by taut cables, and that of the self-stayed support, held in place by special cables only transversely, while the longitudinal stability is guaranteed by the conductors themselves and by the safety cables, or the more expensive version featuring a single stem in steel tubing.[20]

Lightness and economy were to continue to represent the principal objectives in the design of supports for power lines, also when the increased power transported was to call problems of a kind that differed from those concerning the width of the strip of land and the volume of the supports, to the attention of technicians and designers. The consequences resulting from the corona effect (radio interference and noise perceptible to the ear in the surrounding areas), the serious damages caused to living beings by the effects of the electromagnetic field – some of which are ir-

11

12

12. Reinforced concrete poles, patent A. Porr (Vienna), with height variable between 14 and 19 meters.
From: J. Vischer and L. Hilberseimer, *Beton als Gestalter* (Stuttgart, 1928).

13

14

13, 14. Part of the Sluderno-Stelvio-Cesano high-voltage line: welded steel tubing pylons. Nufenen (Switzerland), high-voltage line pylon, "Perfectly successful economy of material and clarity of form [...] steel tubing reinforced with concrete laid and held under tension by steel cables."
From: M. Bill, *Form* (Basel, 1952).

reversible – particularly serious in the presence of the distribution of high levels of tension, are phenomena which have been studied in numerous experimental centers founded more or less everywhere, in Europe and in the United States.

These subjects have given rise to debates between technicians and environmentalists since the 1970s, a decade when, in fact, the European territory and to a large extent the American one having been saturated, with the transition from 380-400 kV in the 1960s to 500 kV through the transmission over long distances in Canada, South America and China, one has witnessed the explosion of the ecological problem. One of the first replies to the many questions raised by the latter has been the launching of programs aimed at removing production stations from the places of consumption. As a direct consequence, the distances over which the energy has to be transported have increased, at the same time as the demand for energy has grown considerably.

The problem of transmission of electrical energy over distances has thus given rise to new design questions, due to the increasing interference caused by the network of lines on the territory. The replies of designers and structuralists have therefore focused on the realization of as compact as possible lines, in order to make the "corridor" narrower; this has led to a preference of lines with relatively small supports placed close to one another.

Also the form of the support has been subject to careful design research, which however has aimed first and foremost at the study of the object as such, considered almost as a theme of an industrial design project, rather than examining the line as a whole and the problems associated with its installation. The result has been singular proposals, such as those presented in America by Henry Dreyfuss in the late 1960s, in an attempt to "beautify" the lines and in particular the supports, which are presented in different versions in wood, steel, aluminum, plastic materials and reinforced concrete.[21] Only some of these proposals have been crowned by success, due to the difficulties of application in a program characterized by radical standardization, but also because contingent situations called for the adoption of alternative solutions that were more advantageous from an economical point of view. One of the more significant parameters in the total cost of the line in fact consists of the manpower. In the first constructions the prevalent solution had consisted of lines with complex supporting structures, through the use of welded tubings that required a high incidence of labor during the construction phase; subsequently one has witnessed a progressive simplification of the trelliswork solutions, featuring ever more frequently separate, bolted elements, aided by a metalwork carpentry, by now organized on the basis of the availability of automatic machines with numeric controls. This tendency has been confirmed in the 1980s, when the importance of the ecological problem was emphasized, at the same time as the use of helicopters – in this, as in many other fields of construction – during the phase of assembly was becoming diffused. It has thus become possible to eliminate the non-specialized manpower in sectors where it has normally been necessary to resort to it when building power lines, especially in particular geographic conditions (savannahs, etc). But while the use of helicopters entails a certain saving in terms of manpower, it follows that preferring helicopters during the assembly phase means to study particular types of trelliswork, whose size is planned on the basis of the carrying capacity of the helicopter.

The use of helicopters also entails an attentive planning of the works and a simplification of the traditional phases of assembly. In the supports for power lines the configuration, the preparation of the elements of the metalwork carpentry, the choice of materials and connections, all therefore represent elements of an integrated design aimed at the best rendition during the phase of assembly and the reduction of total costs to a minimum. In this sense the use of cold-bent sections has proven to be particularly advantageous, in that they make it possible to reduce both the number of poles and the number of joints, thus also lessening the visual impact.[22] One of the interesting consequences of these developments has been the refinement of the regulations in the field of compressed poles and, from a technological point of view, the reduction of the total number of connections, incentivated to use a single bolt per joint, even accepting the a-priori creation of secondary sagging effects.[23,24]

Like the ancient aqueducts, and the road and railway routes, the power lines have, practically already during the years when the problem of transmitting electrical energy over long distances had just been posed, come to confront the subject of the great crossing. The crossings of the St. Lawrence river, of the Bosphorus, of Cadiz, of the Messina straits and, more recently, those of Pearl River in Chine and Gheshm in Iran, have offered as many occasions, counterpointed by solutions in line with

15. F. Depero, "Pali SCAC," advertising placard, 1937.
From: *Fortunato Depero nelle Opere e nella Vita*, ed. by the Legione Trentina (Trent, 1940).

the available technologies of the different periods, to prove the fact that the design problems – especially with regard to the large towers, on many occasions, especially in recent years, built off-shore – are part of the great engineering works. While the studies initially mainly focused on the design research of the conformation of the great supports and on the development of the equipment necessary for the launching of the cables,[25] today the most important factor for the designer consists of the organization of the builder's yard, its calibrated predisposition to accommodate the various phases of the installation.[26]

But while each of these enterprises, considered individually, is characterized by technical and technological considerations of great interest, the network of the power lines eventually create in the land a collection of mute objects, unyielding towards any kind of direct participation, but also ambiguous due to the abstract significance they convey, to which it is increasingly difficult for us to assign a role of identification, of evocative ability, a function of permeability – even indirect – of our imagery.

15

1. The actual nature of the landscape experience can only be clarified by specifying the levels of relation between the formal and biophysical factors which represent the environment and the physical-physiological factors. W. Hellpach, *Geopsiche* (Rome, 1950); Ch. Norberg-Schulz, "Il paesaggio e l'opera dell'uomo," in *Edilizia Moderna*, No. 87-88, p. 63.
2. R. Dubbini, *Geografie dello sguardo. Visione e paesaggio in età moderna* (Turin, 1994). See in particular chapter VI, "Immagini della trasformazione," pp. 140-57.
3. In the mid-1960s Georg Gerster, Swiss photographer and doctor in German literature and philology, proposed, through the taking of air photographs, possible new applications of bird's eye view. His photographs illustrate the different relationships between anthropical and certain parts of nature in which the presence of man seems nonexistent. Cf. G. Gerster, *Der Mensch auf seiner Erde* (Zurich, 1975); *Brot und Salz* (Zurich, 1980); *Flugbilder* (Basel, 1985); *Amber Waves of Grain* (New York, 1990).
4. In the early 20th century the expansion of the electrical industry led, in Europe, to the formation of a heritage of technical-professional and business competences created to a considerable extent ex novo. In Italy in those years the work and inventiveness of certain scientists was determining, especially those of the Polytechnic Schools of Turin and Milan, but also of some less important schools, where the central nucleus of the tech-

nical executives of various size electrical companies received their education. R. Maiocchi, "La ricerca in campo elettrotecnico," in G. Mori (ed.), *Storia dell'industria elettrica in Italia. Le origini. 1882-1914*, vol. 1, p. I (Bari, 1992), pp. 155-99.
5. Cf. R. Giannetti, "Tecnologia ed economia del sistema elettrico," in *Storia dell'industria elettrica* cit., vol. 1, p. I, pp. 355-447.
6. M. Doria, P. Hertner, "L'industria elettrotecnica," in *Storia dell'industria elettrica* cit., vol. 1, p. II, pp. 571-602.
7. L. Geymonat, "Linee generali dello sviluppo delle scienze matematiche e fisico-chimiche nella seconda metà dell'Ottocento," in *Storia del pensiero filosofico e scientifico*, vol. V, pp. 60-91.
8. Some fundamental innovations in the field of electric machinery, such as the "secondary generator" or the transformer of Gaulard and Gibbs, which were to be proven of basic importance for the transmission of electric energy over distances, had been presented at the International Exposition of Turin in 1884 which, as compared to the staging of the "electric prodigies," made a certain sensation, although not comparable to that of the Paris exposition of three years before. P. Meardi, *L'elettricità e le sue applicazioni alla esposizione di Torino del 1884* (Milan, 1885); P. Verole, *La distribuzione elettrica dell'energia e la elettrometallurgia all'esposizione internazionale di Torino* (Turin: Camilla e Bertolero, 1886). It was precisely on the occasion of the Turin exposition that Galileo Ferraris perceived the great possibilities of industrial development of the electric applications.
9. A detailed description of the power station of Santa Radegonda in Milan can be found in G. Colombo, "Illuminazione elettrica," in *Milano, Tecnica dal 1859 al 1884*, ed. by the Collegio degli Ingegneri e Architetti in Milano (Milan, 1895), pp. 459-73.
10. The application of the Edison system in America, England and Germany has been subject to a careful comparative analysis in T.P. Hughes, *Networks of Power, Electrification in Western Society 1880-1930* (Baltimore, 1983), chapters II and III.
11. D. Landes, *Prometeo Liberato. Trasformazioni tecnologiche e sviluppo industriale nell'Europa occidentale dal 1750 ai giorni nostri* (Turin, 1978), pp. 371 ff.
12. A. Banti, "Il trasporto di energia elettrica da Tivoli a Roma," in *L'Elettricista*, I (1892), pp. 281-91, II (1893), pp. 25-31, 49-56, 105-11; "Officine ed Impianti Elettrici della Società Anglo Romana," in *Notizie sui principali impianti elettrici d'Italia*, ed. by the Associazione fra Esercenti Imprese Elettriche in Italia (Milan, 1911), pp. 58-61; C. Cesaroni, *Le centrali elettriche di Tivoli. Il primo trasporto industriale di energia a grande distanza* (Rome, 1937).
13. G. Semenza, "Impianto di Paderno," in *Atti AEI*, I (1898), pp. 121-44; P. Ballerini, *L'impianto idro-elettrico per la trasmissione dell'energia da Paderno d'Adda a Milano* (Monza, 1900), p. 82; "Le rapide di Paderno e la Società Edison," in *L'Elettricità*, XIV (1895), n. 13, pp. 197-8.
14. G. Motta, "L'impianto elettrico di Grosotto in Valtellina a servizio del Comune di Milano," in *L'Industria Elettrica*, I (1911), pp. 12-5; F. Carati, "Sviluppo degli impianti dell'AEM a Milano e in Valtellina dal 1910 al 1950," in *L'Elettrotecnica*, XXXVIII (1951), pp. 219-30.
15. C. Cattaneo, "Strada ferrata da Milano a Venezia," in *Il Politecnico*, vol. IV, 1841, pp. 40-106.
16. "Le plus grandiose project des temps modernes. Abaisser de 200 mètres le niveau de la Méditerranée pour refaire la carte de l'Europe et de l'Afrique," in *Les Echos*, 18 May 1930, p. 26; "Un projet fantastique, l'assèchement partiel de la Méditerranée," in *L'Illustration*, n. 4490, 23 March 1929; H. Sörgel, *Atlantropa* (Zurich-Munich, 1932). An accurate analysis of the social-political significance of the Atlantropa project can be found in W. Voigt, "Le plus grandiose projet des temps modernes," in J.-L Cohen and H. Damisch (ed. by), *Américanisme et modernité* (Paris, 1993).
17. R. Huhn, "Art et technique: la lumière," in *Cinquantenaire de l'Exposition internationale des arts et des techniques dans la vie moderne* (Paris: IFA, Paris-Musées, 1987), pp. 392-403.
18. Cf. the introduction, ed. by H. Dreyfuss, of the monograph *Electric Transmission Structures, a Design Research Program* (New York: Electric Research Council, 1968), pp. 9-11.
19. In Italy, for example, the gradual creation – during the first decade of the 20th century – of a national electric system, with the relative infrastructures and the new restrictions thus created, was the result of a laborious research which had enabled the electric industry to successfully overcome the long years of World War I and the tormented vicissitudes which were to characterize the years after the war in Italy until the advent of Mussolini's dictatorship (1925). The characters of localism which had characterized the prewar development had given way to the creation of an aggregate of regional electric systems and, in some cases, as in continental Southern Italy, multi-regional. The creation of these multi-regional systems for the use of electric energy, in some cases produced far from the place of consumption, was, at the same time, cause and effect of a process of financial and technical concentration which, in the decade from 1915 to 1925, had favored the creation of some great holding companies, which had contributed to the gradual creation of the land systems. R. Giannetti, "Vecchi e nuovi sistemi territoriali," in L. De Rosa (ed. by), *Storia dell'industria elettrica in Italia. Il potenziamento tecnico e finanziario. 1914-1925*, vol. 2 (Bari, 1993), pp. 235-316. The appearance of legislative problems concerning the right of way of power lines are marked in Italy by the approval by Parliament (1894) of the important law on the right of way of power lines, which had the objective of facilitating the passage of power lines, suspended or underground, for the transmission of electrical energy through private properties.
20. G. Magenta, "Realizzazioni e progetti italiani 1963-64 di pali e antenne," in "Atti delle Giornate Italiane della Costruzione in Acciaio," Pisa, 11-12 September 1964, published in *Costruzioni Metalliche*, No. 1, 1965, pp. 1-6; M. Comolo, "Impiego delle strutture strallate nelle linee elettriche," in *Costruzioni Metalliche*, No. 1, 1965, pp. 40-4; S. Iudica, A. Spinelli, "Struttura strallata per l'esecuzione di prove dielettriche. Progetto 1000 kV," in *Costruzioni Metalliche*, No. 5, 1984, pp. 291-8; A.R. Flint, "Towers and Masts," in *The Art and Practice of Structural Design* (London: The Institution of Structural Engineers, 1984), pp. 120-4; C. Avril, *Construction des Lignes Aériennes à Haute Tension* (Paris, 1974).
21. H. Dreyfuss, *Electric Transmission* cit.
22. A. Catenacci, P. Faggiano, "Nuove tipologie di sostegni in profili piegati a freddo per linee elettriche a 500 kV: la linea sperimentale dell'Ente di Ricerche Elettriche Americano (EPRI)," in *Costruzioni Metalliche*, No. 3, 1987, pp. 144-52.
23. *ECCS, Technical Committee 8, Structural Stability. Technical Working Group 8.1.* "Computer Aided Design Programmes for Latticed Towers," 1st ed., 1987, No. 51.
24. I am grateful to Engineer Massimo Finzi, chairman of the Associazione Costruttori in Acciaio Italiani and SAE Towers (Milan) for his kindness and collaboration in providing me with useful information and outlining, from a technical point of view, the main phases of the process of transformation of the design problems concerning power lines. Particular thanks to ENEL spa, Turin Branch, and the Azienda Energetica Municipale of Turin for having granted me access to the images and documents conserved in their respective archives.
25. Società Generale Elettrica della Sicilia, *L'attraversamento elettrico dello stretto di Messina* (Rome, 1958); A.M. Toscano, "Le Torri di Messina," in *Costruzioni Metalliche*, No. 5, 1960, pp. 205-25; of the same author cf. "Le Torri di Cadice," in *Costruzioni Metalliche*, No. 4, 1961, pp. 182-94.
26. M. Finzi, S. Iudica, M. Leva, E. Portaluppi, "Elettrodotto 500 kV per l'attraversamento del Pearl River in Cina," in *Costruzioni Metalliche*, No. 5, 1988, pp. 207-1; M. Leva, B. Salesi, "Attraversamento a 230 kV dello stretto di Gheshm con impiego di struttura off-shore," in *Costruzioni Metalliche*, No. 3, 1994, pp. 13-24.

16. H. Sauvage, transformation sub-station at the Exposition des Arts Décoratifs, Paris, 1925. Sculptures by Z. Sauvage. From: Fondation Electricité de France/ IFA, *Architecture de l'électricité* (Paris: Norma, 1992).

16

17. Ternary splitting on "cat"-type supports in zones with high overload. ENEL Archive, Turin Branch.

17

News from the companies

Aletto and Canaletto

The technology of sleep
Design by Paolo Piva
and Antonio Citterio
B&B Italia

One area in which the term "ergonomics" is of greater significance than ever is that of sleep: in view of the fact that people spend a third of their lives asleep, the importance of the bed and the elements of which it is composed - bedspring, mattress, pillows - seems obvious. B&B Italia, in line with its tradition of research and innovation, has produced a series of beds designed by Paolo Piva and Antonio Citterio, in which the quest for new aesthetic solutions is combined with the best sleep "technology".

Aletto

An image of essential force and technological rigor, for the rarefied climate that is a mark of Aletto, designed by Paolo Piva. Two wings of cast metal arch from the soft fabric bed spring that protects the back during flights of dream. In its simplicity, the sober brilliance of aluminum is combined with the dense, though light and transparent, mesh of the cotton. The absolute decision of the only version available permits only a single choice, between black and beige. But to fit in with one's nighttime habits, there is also the possibility of mounting a rotating bedside table, that is practical, discreet, and silent. In a happy marriage, the cold rib of cast aluminum stretches the warm and enfolding cotton head-board and softens it with the yielding roll, made of fabric or leather, that serves as a loins rest.

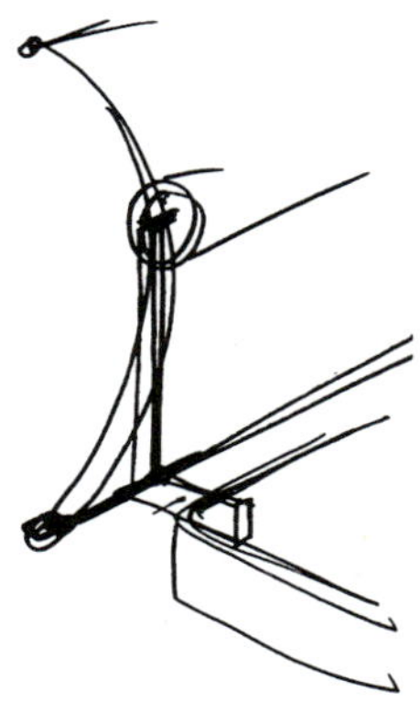

The sober, slightly sinuous lines define a soft and minimal simplicity. The secret of Canaletto, designed by Antonio Citterio, lies in the curve between head and side which, perhaps in homage to the name of the Venetian painter, stirs memories of traditional gondola beds. The hood of the back slips between mattress and head-board in perfect continuity.

The visual effect that this creates is one of a soft harmony among the coordinated series of fabric coverings. And if one wishes... to uncover all the elegance of these beds, their coverings are all interchangeable. Various solutions for the supporting feet as well, covered in leather or fabric, with cover stretching to the ground or mounted on casters.

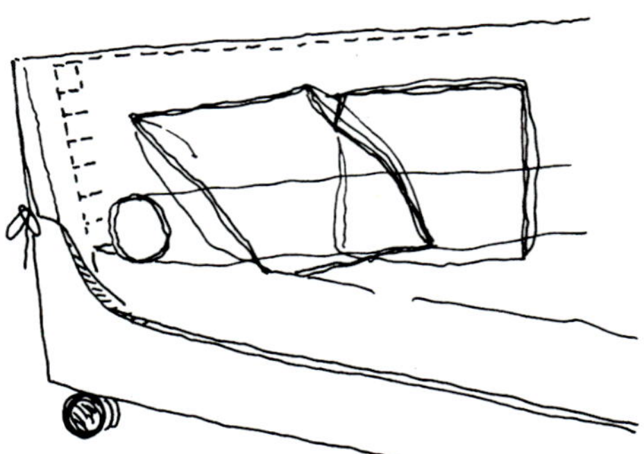

Castelli for The Italian Territory

Three-dimensionality at the service of art
Castelli

1

photo Beppe Raso

In September Castelli offered their customers a new event; an event that has offered concrete proof of the company's desire to confirm their identity as a firm with an interest in culture and innovation.
In the pleasant setting of the Museo dei Navigli in Milan, an exhibition has been prepared called "Territorio Italiano" (Italian Territory), leading to an area created completely using Castelli's new 3D office system. This product proves it can

3

2

resolve the traditional antinomy between flexibility and simplicity, lending itself, thanks to the use of the small number of elements of which it is composed, to redesigning a Renaissance plan so dear to Italian architectonic culture.
Cupboards/towers looking onto writing desks/gardens, reception desks/walls hiding the secrets of worktops/palaces. Thus, going through the wooden door, symbolizing since 1877 the Castelli cabinet maker's shop, we find ourselves in a dream atmosphere where age-old tradition merges with state-of-the-art technology in that synthesis of form and structure that has brought Castelli so much success.
With the help of the most sophisticated electronic technology and with an entertaining script to work with, the company has created

4

a film in which we can appreciate the modular character of the 3D system, the line and the colors that make it perfect in every context: from the huge multi-purpose loft to the prestige office, from the meeting halls to the traditional rooms. All this was the setting prepared by Castelli for the "Territorio Italiano" exhibition.
In 1993 Giacinto di Pietrantonio asked a number of contemporary artists, all used to working with three-dimensionality, to take up a stimulating challenge: the interpretation of the Italian landscape. The philosophy of the project is to once again offer the ideal conditions that made Italian art so great in history, thus offering this extraordinary country as the setting/inspiration for the artists' imagination.

So it is that 'Territorio Italiano" has come about – more than an exhibition, a message that is universal and, at times, controversial, intended to acquaint the public with the state of contemporary art. By proposing this new edition of the exhibition, enriched by other important works, Castelli confirm their aptitude for originality and innovation whilst fully respecting tradition.

5

On the preceding pages:

1, 3/5. Different layouts of the 3D system.

2. The wooden door symbolizing Castelli since 1877.

6. Doctor Alberto Ponzellini, chairman of Castelli, with some guests.

7, 9, 11, 12. Some guests during the event.

6

7

8

9

8. A room for audiovisuals with Templum furniture.

10. General view of the "Territorio Italiano" exhibition.

10

11

12

Dodicesima Notte

The first all-natural system of beds
Design by Luca Meda
Molteni & C.

A new awareness of healthy sleeping has come about. The best layout for the bedroom now includes support structures and bases made of anatomical wooden struts, mattresses and pillows with natural padding, bed linen made of ecological fabrics. The bio-compatible bed must be constructed limiting the number of metal pieces as far as possible to avoid the negative interferences of electromagnetic flows in the body. At Molteni, they have always been convinced of the importance to be attributed to the bedroom: a prime example of this is the "Morna" bed that introduced the all-wood concept in 1971 for all parts including the support that was traditionally made of metal. With Dodicesima Notte (Twelfth Night) designed by Luca Meda, the functional sleeping tradition and the adoption of a sound design criterion is no longer limited to the bed but extends to the surrounds as well, creating a system where fitted headboards, sliding platforms, modular wall and other components create a comfortable bedroom expressing the personal style of the users. Each piece can be added over time to the basic structure for versatile use and gradual completion of the furniture, it is enough

Dodicesima Notte is a system of beds made of multi-layer atoxic wood with flexible beechwood struts, heartwood accessories, headboards that breathe and just a few essential metal joints.

just to place the bed over the platform, a new bedside table on the fitted surface or add a module with shelves on the wall. The heartwood sections of the beds and furnishings use essences of beech and walnut, the multi-layer sections use panels which come from specific locations without formaldehyde and assembled with atoxic glues; padding is in coco, latex foam and cotton, the fabrics are pure cotton and mixed linen. For an environment where the night is pleasant and sleep healthy, and where the day is just as relaxing, reading, writing or telephoning a friend.

The Flaminio Business Center in Rimini

Four external panoramic elevators
Design by S. Piccioli and G. Perazzini
Sabiem

The Flaminio Business and Commerce Center, which was recently inaugurated, represents a significant step ahead in terms of enhancing the urban territory of Rimini; it is located along the ancient via Flaminia.
The decision to build a high quality complex, archetype of a future evolution of the real estate sector, has above all resulted in an original architectural form. The designers have thus adopted unusual volumes, forms and materials: the building is characterized by its vertical load-bearing structure (entirely in steel, from its foundations), the casing comprising mirror façades and curved aluminum elements and, most importantly, the four external panoramic elevators by Sabiem.
The construction details of the panoramic elevators have been subject to particular attention. The shaft is located on the outside of the building, without the customary transparent covering. All connecting elements have therefore been zink-plated to prevent corrosion, and new doors have been studied for the shaft to meet the particular requirements in terms of functionality and regulations.
The openings are in fact not only waterproof, to avoid the penetration of rainwater; they also feature fire-resistant doors (Rei 120) even though they are positioned externally.
Each elevator carries up to 13 persons, who travel comfortably, admiring the splendid surrounding landscape (hills and the sea, with the Republic of San Marino in the background) at a speed of 1.6 meters per second and up to a height of 30 meters.
The Flaminio Center, built by Rimini-based Ge.Cos., provides the capital of the Romagna region with a high-quality complex featuring truly advanced constructive and technological solutions.

Massimiliano Colombo

Living Tomorrow

The House of the Future
Project by Frank
and Maurice Beliën
Abet Laminati

We live in an ever changing world. Our society is evolving faster and more fundamentally than ever. Our demands for the quality of our domestic and living environments are constantly increasing. Man no longer wants to continue polluting the environment. Better waste management, energy saving and the use of environmental-friendly materials are now important issues. In the House of the Future one wishes to live in accordance with this new era.
In our future home we want to dispose of home-shopping, telephone banking and a home-office. We would also like to see our domestic and living environments become safer, healthier and more flexible in the future.
If we want to live in this House of the Future, we must start work on it today.
In 1989, architect Frank Beliën developed, together with his father architect Maurice Beliën, the idea to evoke a future living environment, while searching to answer basic questions as "How do we work in future?", "How do we travel, live, eat, sleep, etc... in future?"
Therefore, leading companies in different branches were contacted in order to participate in the project, on condition that delivered materials or services were innovative. Nowadays the project welcomes already over 100 participating companies.
The 5th of October 1993, the foundation-stone was officially laid, and on the 16th of March 1995, the official inauguration of the Living Tomorrow complex has taken place.
Living Tomorrow is located at Vilvoorde, near Brussels, on a total 15,000 m2 surface, 1120 of which are used for the experimental house, and is composed of 4 different zones: Office of the Future; House of the Future; Information Complex with an auditorium and exposing facilities; Lodges at the disposal of the participating companies.
Although Architect Bureau Beliën has to be considered as the initial entrepreneur, there are over l00 participating companies and co-entrepreneurs, among whom are to be mentioned the Flemish Community and Tractebel, and Abet Laminati.
Abet Laminati identified from the beginning their own continuously research for innovative laminate products with the well-based initiative "Living Tomorrow".

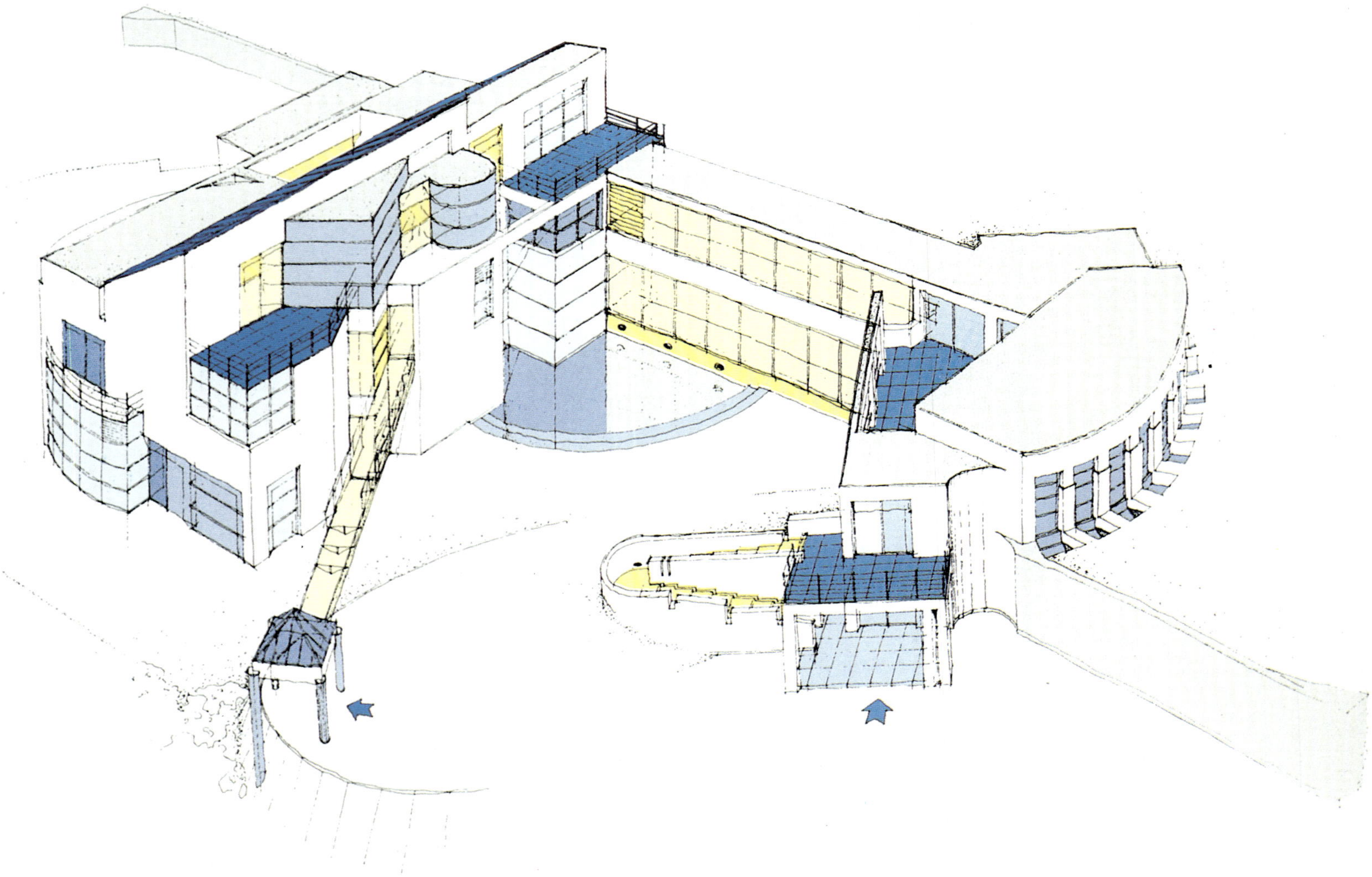

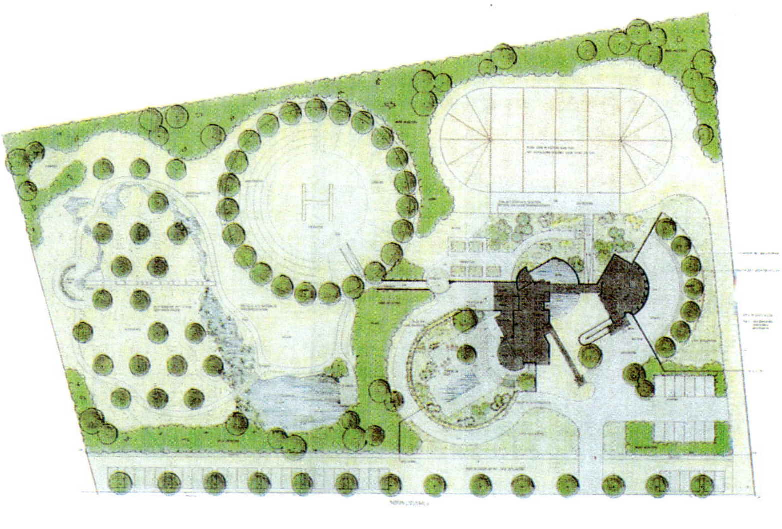

Abet Laminat applied following HPL-products:

1. MEG Standard and MEG Architettura outside wall-covering HPL panels;

2. HPL Flooring Parqcolor;

3. Print HPL type "Metalli";

4. Furniture designed by Ettore Sottsass "Mobile Bar".

100 1. MEG Standard and MEG Architettura:
Architect Frank Beliën created an own design to be printed on the basic MEG material within a special department of the Abet LAminati plant in Bra (Italy). The MEG wall panels are glued upon a wooden frame, using the most advanced glueing techniques. The standard MEG wall panels are cut in different sizes and patterns in order to compose with differents pieces and different colors, the foundations of the Living Tomorrow building.
MEG Standard Grey 871 and 475
MEG Architettura Brown 799
Dimension: 3050 x 1300 x 6 mm
Architect Frank Beliën: "MEG is a genius wall covering product as there are almost no limits towards prints and fixing patterns. The invisible fixing method by means of glue gives a perfect result, as well as on a technical as on an aesthetical level."
Total surface: 40 M2
In addition to these customized MEG panels, an additionnal quantity of 120 M2 is applied as a covering of 5 supporting pillars. The MEG panels are installed on a wooden frame by means of self-centring screws, paint in the same color as the panels.
MEG Architettura Brown 799
Dimension: 3050 x 1300 x 6 mm
Architect Frank Beliën:
"The supporting-pillars, made out of silicate stone filled with concrete, caused some aesthetical troubles. The MEG panels brought the solution because of their easy way of installing and their various available colors."

2. HPL Flooring Parqcolor is a tongued-and-grooved floating floor, 6.5 mm thick, composed of a panel MDF bonded between Print HPL flooring grade laminate, and is highly resistant to wear. Parqcolor 9226 "Beech" in the children's room, 9926 "I Tronchi" in the Home Office, 9937 "I Tronchi" in the lodges' bar. Parqcolor is an innovative product in many ways: the connection to other flooring types (linoleum and carpet) is made with fine flexible connection-profiles. In the lodges' bar, it is integrated in a stone tile pattern and connected by means of a silicon joint. The almost unlimited laying possibilities together with the various available wood designs and wood reproductions, the high performance (high resistance to wear, scratching etc.) and its environmental-friendly aspect make it usable for different applications (domestic and semi-industrial).

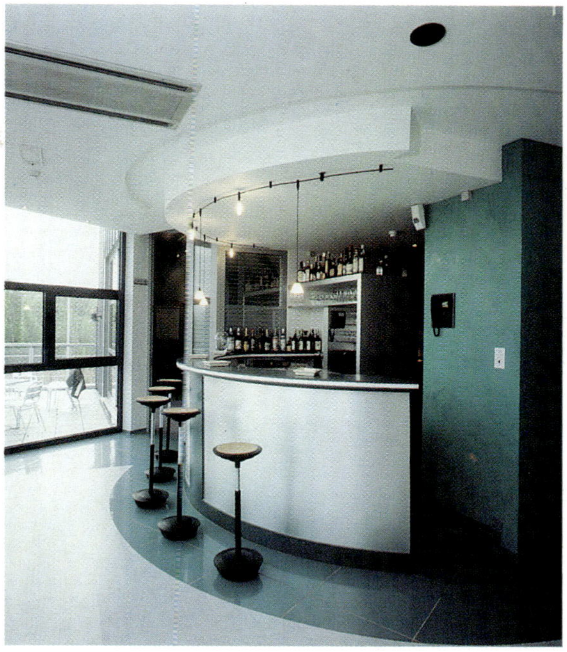

3. Print HPL type "Metalli": references 877 Satinata and 877 Millerighe have been used in the design of furniture.
According to Beliën, the "high-tech look" of the Print "Metalli" laminate goes perfectly together with the sofisticated equipment all around in the complex, even in the bathroom. Besides the contrast of the hard and cold Print "Metalli" laminate with the warm and soft tiles is completely compatible with the diverse contrasts such as warm-cold; domestic-industrial; quiet-busy, everywhere appearing in the complex. Moreover, the conception of aluminum-like furniture was never that easy before thanks to Print Metalli laminate.

4. "Mobile Bar" designed by Ettore Sottsass: a special Print "Seriegrafia" laminate was produced in order to manufacture this piece of furniture.

Johnny Maris

Light for Art

A program to illuminate the great masterpieces of Italian art
ENEL

1. Room in the Vatican Art Gallery
2. Basilica of Saint Francis in Assisi.

Light for Art is the program promoted by ENEL as proof that their presence on the Italian scene is not just entrepreneurial but also cultural. The extensive background of technology and experience gained by the Company over years of specific activity is made available to the public through the designing and construction of sophisticated lighting installations, to bring out of the dark some of the masterpieces of the largest museum of art and architecture in the world: Italy.
Yet Light for Art is not just light. It is an extensive cultural and informative project, combining the preparation of systems based on special lighting projects with a series of important interrelated initiatives, such as the realization of studies and photographic documentation of the works examined, the publication of art books and the production of multimedia instruments such as CD-ROM and virtual reality programs, using the most sophisticated Information Technology, always within a rigorously scientific framework.
Light for Art is therefore a complete, highly structured package with extensive technological and communicative content which can be used by all the institutional, economic and productive structures that intend to contribute to the enhancement of a monument or a work of art in its own local setting, through actions to improve access to it and to highlight its content.
The initiative, conducted in coordination with the regional branches of the Monuments and Fine Arts Service responsible for identifying the various works of art to be illuminated and in tune with the orientations of the Heritage and Environment Ministry, was launched in April 1990 with the lighting for the mosaics of the Basilica of Saint Mark in Venice.
The spectacular results of this first project have led to the full appreciation of the marbles, the columns and above all of the complex of mosaics that were hidden in the half-light of the great arcades.
The Light for Art program then proceeded, in September 1990, with the Cathedral

foto Pianeta Immagine

1

3

3, 4. Lighting Modules.

in Chieri, one of the most important examples of Piedmontese Gothic.
Then it was the turn of the Cathedral of Gemona, the symbol of the decade-long, active reconstruction work following the earthquake of 1976, which was lit up in January 1991. During the same year projects were carried out involving Piazza del Plebiscito in Ancona, the Basilica of Saint Vitale and the Mausoleum of Galla Placidia in Ravenna, and the Upper and Lower Basilicas in Assisi. Between 1992 and '93 the lighting systems designed for the Piazza and the interior of the Cathedral of Spoleto, the Vatican Art Gallery, the Basilica of Saint Laurence in Florence and the Rooms frescoed by Raphael in the Papal Rooms of Julius II and Leo X in the Vatican were all inaugurated. Crowning these are the projects for spotlighting, that is, short-term lighting, realized using Lighting Modules – a technologically advanced, innovative system used for artistic or cultural events. Examples of this are the illumination of the Maschio Angioino, of Castel dell'Ovo, of Sant'Elmo in Naples and the splendid monumental complex of the Reggia in Caserta, for the G7 meeting; of the Imperial Forums and the Trajanean Markets, or again, recently, the illumination of the Cathedral and Cathedral Square in Lecce and the Shrine of Loreto, for the visit by the Holy Father.
The guiding motif for all these projects, besides the obvious need to improve access to and the enjoyment of the individual masterpieces through the diversification of lighting installations, has always been that of fully respecting the atmosphere surrounding the work; that is, not flattening its perspective, not bringing out its volumes too scenographically and, above all, not altering the balance of the effects of natural light as originally perceived by the author. Particular attention has been devoted to containing the consumption of electrical energy, an objective achieved by the appropriate and rational management of technology, within a framework of maximum efficiency.

4

5

5. Art volumes published by ENEL.
6. G.L. Bernini, *Enea, Anchise e Ascanio.* Galleria Borghese, Roma.

As has been said, there have been numerous initiatives connected with the program of design and lighting technology in the strictest sense, all aimed at the enhancement, dissemination and creation of a deeper awareness of Italian artistic heritage.
In the publishing field, mention must be made of the numerous art books providing evidence, through detailed photographic documentation, of the technical interventions made and the artistic wealth of the sites involved. Among these, on the occasion of the illumination of the Basilica of Saint Mark in Venice, two prestigious volumes were published. The successful lighting technology project in the Basilicas of Saint Francis in Assisi was also commemorated by the publication of an important book. Also of fundamental importance are the volumes linked to the operation carried out on the Vatican Art Gallery, and on the Rooms frescoed by Raphael in the Vatican. The latest publishing events are devoted to the Galleria Borghese and the Imperial Forums. The videos and multivisuals connected with the most significant projects must also not be forgotten, such as the ones made about the frescoes of Saint Francis in Assisi, explored in their multiple perspectives.
Light for Art also avails itself of the most advanced information technology.
Of the different technologies, the most significant is undoubtedly virtual reality, which is able electronically to recreate settings and situations, involving the spectator with effects of astonishing realism.
An example of this is "Giotto's City", a program through which the visitor can "enter" the frescoes by the Master in the Basilica of Saint Francis of Assisi and wander around a non-existent city, rebuilt based on the images painted by the artist.
Equally evocative is the trip into the Basilica of Saint Peter, carried out following along the marvellous perspective of the colonnade into the church, with the possibility of visiting the old pre-existing Constantinian central core, rebuilt based on careful scientific studies. These applications use an extremely sophisticated computerized system, the result of long and complex research by ENEL.
Very important is also the realization of a series of CD-ROMs produced by ENEL, intended to disseminate art intelligently, through special programs for visits to exhibitions and museums, and for the simulated interactive restoration of famous statues and paintings.
Thus humanistic culture and technology have merged in realizations of great significance. Over the years Light for Art has become an authoritative promoter and communicator of culture.

6

Dove trovare Rassegna

Librerie

- **Asolo**
  Libreria Polo Cart.
- **Bari**
  Libreria Feltrinelli
- **Bologna**
  Libreria Feltrinelli
  Libreria Il Leonardo
- **Bolzano**
  Libreria K.O. libri
- **Catania**
  Libreria de Martiniis
- **Cividale del Friuli**
  Libreria Muner G.
- **Como**
  Ass. culturale
- **Conegliano**
  Quartiere Latino Libri
- **Ferrara**
  Libreria Feltrinelli
  Architecnica
- **Firenze**
  Coop. libraria universitaria
  Coop. Un. Studio Lavoro
  Libreria Alfani
  Libreria Centro d'Arte Spazio Tempo
  Libreria Feltrinelli
  Libreria Lef
  Libreria Marzocco
- **Genova**
  Libreria Feltrinelli
- **Livorno**
  Libreria Belforte
- **Mestre**
  Libreria Don Chisciotte
- **Milano**
  Libreria Cortina
  Libreria CLUP
  Libreria Cusl
  Libreria Feltrinelli (Baires)
  Libreria Feltrinelli (Europa)
  Libreria Feltrinelli (Manzoni)
  Libreria Hoepli
  Libreria L'Archivolto
  Libreria M. Sedis (C.so Garibaldi)
  Libreria M. Sedis (Galleria)
  Libreria Unicopli
- **Modena**
  Libreria Feltrinelli
  Libreria M. Sedis
- **Napoli**
  Libreria CLEAN
  Libreria Guida
  Libreria Feltrinelli
  Libreria Il Punto
- **Padova**
  Libreria Feltrinelli
  Libreria Ginnasio
  Libreria Progetto
- **Palermo**
  Libreria Dante
  Libreria Feltrinelli
- **Parma**
  Libreria Feltrinelli
- **Pavia**
  La Libreria
- **Pescara**
  Libreria dell'Università
  Libreria Feltrinelli
  Libreria Filogrosso
- **Pisa**
  Libreria Feltrinelli
- **Pordenone**
  Libreria Al Segno
  Libreria Il Labirinto
  Libreria La Rivisteria
- **Ravenna**
  Libreria La Rinascita
- **Reggio Calabria**
  Libreria Pepo
- **Reggio Emilia**
  Libreria Vecchia Reggio
- **Roma**
  Libreria Dedalo
  Libreria Feltrinelli (Argentina)
  Libreria Feltrinelli (Babuino)
  Libreria Feltrinelli (Orlando)
  Libreria Kappa
  Libreria Rinascita
- **Rovereto**
  Coop. Libraria Rovereto
- **Salerno**
  Libreria Feltrinelli
- **Siena**
  Libreria Feltrinelli
- **Taranto**
  Libreria Dichens
- **Torino**
  Libreria Agorà
  Libreria Celid
  Libreria Comunardi
  Libreria Feltrinelli
  Libreria Internazionale Salone
- **Trento**
  La Rivisteria
  Libreria Universitaria
- **Treviso**
  Galleria del Librario
  Libreria Canova
  Libreria Marton
- **Trieste**
  Coop. Fra Servi
  Libreria Italo Svevo
  Libreria La Fenice
  Libreria Tergeste
- **Udine**
  Libreria Tarantola
- **Venezia**
  Libreria Arsenale
  Libreria Nuova Cluva
  Libreria Fantoni
- **Verona**
  Libreria Catullo
  Libreria Rinascita
- **Vicenza**
  Libreria Galla
  Libreria Spazio Più
  Libreria Traverso

- **Lugano**
  Libreria Melisa
- **Madrid**
  Xarait Libros

Distributori

- **Joo Distribuzione**
  Via F. Argelati 35, 20123 Milano
  tel. 02/8375671
- **Quadrifoglio Libri** (Tre Venezie)
  Via S. Breda 24/B, 35010 Limena, PD
  tel. 049/768099 - 8840276

15 Rassegna (Walter Gropius 1907/1934)
21 Rassegna (Louis I. Kahn 1901/1974)
25 Rassegna (Siegfried Giedion: un progetto storico/Siegfried Giedion: A History Project)
30 Rassegna (Piet Zwart: L'opera tipografica 1923-1933/Piet Zwart: The Typographical Work 1923-1933)
31 Rassegna (Interni a Milano e a Como 1927-1936/Interiors in Milan and Como 1927-1936)

No. 64 (Edinburgh)
edited by Robert Tavernor

The urban history of Edinburgh can be schematically distinguished in two phases: the ancient city, the "Old Town", and the "New Town"; a small valley separates the old settlement from the city erected during the Age of Enlightenment. In 1766 the municipal Council announced a competition for the idea of a new town. The plan presented by the winner, James Craig, consists of a simple, regular grid. Two squares are united by a central street, which is flanked by two parallel streets. A large area was therefore planned in accordance with a rigorous architectural scheme. Subsequent constructions designed by Robert Adam contributed to the further definition of the new town. Edinburgh grew once again in the 19th century, earning the name of Athens of the North. It is very significant that Patrick Geddes chose to present the case of Edinburgh at the RIBA exhibition held in 1910, one of the fundamental events of modern town planning, to develop his method of urban research.

---

65 (Avant-garde Architecture in Poland 1918-1939)
edited by Adam Milobedzki

Very little is known about new architecture in Poland; even recent historiography merely remembers the figures of Szymon and Helena Syrkus, who were very active in the late 1920s, also on the international stage. Szymon Syrkus participated in the foundation of Ciam in 1928 at La Sarraz and was to use this organization once more in 1933, to present and discuss his projects for "functional Warsaw", a plan on a regional scale. While these two figures represent the point of contact between the Polish architectural avant-garde and the ideas of new European architecture, the internal vicissitudes of this movement are no less interesting. It is sufficient to mention the magazine *Blok*, first published in 1924, the organ of the Polish constructivist movement and characterized by a typographic research of considerable importance; or the activity of discussion of theoretic research of the group "Praesens," also in the 1920s.

EDITRICE
COMPOSITORI

RASSEGNA

CIPIA s.r.l.
Via Stalingrado, 97/2°
40128 Bologna
I T A L Y

Aletto et Canaletto

La technologie du sommeil
Projets de Paolo Piva et Antonio Citterio
B&B Italia

Un domaine où le terme "ergonomie" a plus que jamais une raison d'être est celui qui est lié au sommeil : si l'on pense que l'homme passe un tiers de son existence à dormir, l'importance que revêt le lit et les éléments qui le composent, à savoir sommier, matelas, oreillers, devient évidente.
B&B Italia, fidèle à sa tradition de recherche et d'innovation, a produit une série de lits dessinés par Paolo Piva et Antonio Citterio, où la recherche de nouvelles solutions esthétiques est associée à la meilleure "technologie" du sommeil.

Aletto

Une image de force essentielle et de rigueur technologique, pour le climat raréfié qui distingue Aletto, dessiné par Paolo Piva. Deux ailes métalliques moulées cambrent le filet textile moëlleux qui protège le rêve. Avec simplicité, le brillant sobre de l'aluminium se marie à la trame serrée, quoique légère et transparente, du coton. La décision absolue d'une unique version disponible ne s'accorde qu'une seule variante, le noir et le beige. Mais pour s'accorder à toutes habitudes nocturnes, il reste toujours la possilité de monter sur la structure une tablette pivotante, utile, discrète et silencieuse.
En un mariage réussi, le nerf, froid, moulé en aluminium tend la tête en coton, chaude et enveloppante, et l'adoucit avec le traversin moëlleux, en tissus ou en cuir, fixé en guise d'appui-reins.

Canaletto

Sa ligne sobre, légèrement sinueuse, dessine une simplicité fine et minimale. Le secret de Canaletto, dessiné par Antonio Citterio, tient dans la courbe entre la tête et le cadre, qui, peut-être en hommage au nom du peintre vénitien, renferme la mémoire des lits en gondole traditionnels.
La housse du dossier s'insinue entre le matelas et la tête en une continuité parfaite. L'impact visuel qui se crée est celui d'une douce harmonie entre les séquences coordonnées des revêtements textiles. Et si vous voulez... déhousser toute l'élégance de ces lits, leurs revêtements sont tous interchangeables. De nombreuses solutions sont également disponibles pour la base : pied recouvert de cuir ou de tissu, avec housse jusqu'à terre ou sur roulettes.

Castelli pour le Territoire Italien

La tridimensionnalité au service de l'art
Castelli

En septembre dernier, Castelli a offert à ses clients un événement nouveau qui a marqué d'un signe concret la volonté de confirmer sa propre identité d'entreprise attentive à la culture et à l'innovation. Dans le cadre agréable du Musée des Navigli de Milan, il a été organisé une exposition intitulée "Territoire Italien" qui menait à une zone complètement réalisée avec le nouveau système pour bureau 30 signé Castelli. Ce produit démontre qu'il est possible de résoudre l'incompatibilité traditionnelle existant entre la souplesse et la simplicité, en se prêtant à retracer, à l'aide du très petit nombre d'éléments qui la composent, un plan de style Renaissance si cher à la culture architecturale italienne.
Des armoires/donjons qui se referment sur des bureaux/jardin, des meubles de réception/murailles qui cachent les secrets des plans de travail/palais. Ainsi, en franchissant la porte en bois qui est, depuis 1877, le symbole de l'ancienne ébénisterie Castelli, on se retrouve dans une atmosphère de rêve où la tradition antique se mélange à la technologie d'avant-garde, en une synthèse des formes et des structures qui a conduit Castelli sur la voie du succès.
Grâce à la technologie électronique la plus sophistiquée et à une mise en scène amusante, l'entreprise a réalisé un film qui permet d'apprécier la modularité du système 3D, la ligne et les couleurs qui le rendent parfait dans n'importe quel cadre : de la grande salle multifonctions au bureau de représentation, de la salle de réunions aux milieux traditionnels.
Tout cela se trouvait dans le cadre que Castelli avait préparé pour l'exposition "Territoire Italien".
En 1993, Giacinto di Pietrantonio demanda à de nombreux artistes contemporains, habitués à travailler dans le secteur de la tridimensionnalité, de coopérer à une entreprise passionnante : l'interprétation du paysage italien. La philosophie du projet est de reproduire les conditions idéales qui ont rendu l'art italien si grand au cours de l'histoire, livrant ainsi cet extraordinaire pays, en guise de scénario/inspirateur, à l'imagination des artistes.
C'est ainsi que naît "Territoire Italien", plus qu'une exposition, un message universel et frisant parfois même la transgression, pour faire connaître au public l'état de l'art contemporain.
En proposant cette nouvelle édition de l'exposition que l'on a enrichie de nouvelles oeuvres importantes, Castelli confirme sa tendance à l'originalité et à l'innovation, dans le plus grand respect de la tradition.

Dodicesima Notte

Le premier programme pour la nuit totalement naturel
Projet de Luca Meda
Molteni & C.

Dormir sainement est devenu désormais une exigence. A l'heure actuelle, organiser correctement l'espace nuit signifie proposer des structures de supports et des sommiers à lattes anatomiques en bois, des matelas et des oreillers rembourrés de fibres naturelles, du linge de lit en tissus écologiques. Le lit bio-compatible doit être construit en limitant au maximum l'utilisation de parties métalliques qui génèrent des champs électromagnétiques interférant avec les rythmes naturels du corps. Molteni a toujours accordé un soin particulier à l'espace nuit : le lit "Morna" l'a démontré en introduisant, en 1971, le concept du "tout-en-bois" également pour la structure, proposée traditionnellement en métal. Avec Dodicesima Notte, conçu par Luca Meda, l'expression de la tradition, de la fonctionnalité et du design ne se limite pas simplement au lit mais se mêle également à son environnement, en proposant un programme élaboré où têtes de lit équipées, estrades coulissantes, composants modulaires de parois ou autres permettent de réaliser une chambre personnalisée et confortable. Les éléments se rajoutent à la structure et permettent une utilisation universelle et une réalisation progressive de l'ameublement : il suffit de superposer le lit à l'estrade, une nouvelle table de nuit à la surface équipée ou rajouter un module avec étagère à la paroi. Les parties en bois massif des lits et des accessoires d'ameublement sont en hêtre et en noyer, celles en multiplies utilisent des panneaux d'origine contrôlée, garantis sans formaldéhyde et assemblés avec colles non toxiques. Les rembourrages sont en coco, latex et coton, les tissus en pur coton et toile métissée. Tout ceci dans la perspective de réaliser et garantir un espace où la nuit soit agréable et le sommeil salutaire et, le jour, il soit possible d'y séjourner confortablement pour lire, écrire ou téléphoner à un ami.

Le Centre de Direction Flaminio à Rimini

Poker d'ascenseurs panoramiques externes
Projet de S. Piccioli et G. Perazzini
Sabiem

Le Centre de direction et commercial Flaminio, récemment inauguré, représente une importante intervention de requalification du territoire urbain situé le long de l'ancienne Via Flaminia, à Rimini.
Le choix de la construction d'un édifice de qualité, archétype d'une évolution future du produit dans le domaine du bâtiment, a tout d'abord donné naissance à une forme architecturale originale. C'est pourquoi les architectes ont utilisé des volumes, des formes et des matériaux insolites : l'édifice est caractérisé par une structure portante verticale (entièrement en acier, à partir même des fondations), par un revêtement externe fait de façades réfléchissantes et d'éléments courbes en aluminium et, surtout, par les quatres ascenseurs panoramiques externes Sabiem.
Une attention particulière a été accordée aux détails de construction des ascenseurs panoramiques, dont la cage est entièrement extérieure, sans la protection des habituels éléments transparents. Par conséquent, tous les points de raccordement ont été zingués pour éviter les phénomènes de corrosion, et de nouvelles portes d'étage ont été

étudiées pour satisfaire aux caractéristiques fonctionnelles ainsi qu'aux normes requises. En effet, les ouvertures sont étanches pour empêcher la pénétration de la pluie et sont également équipées de portes résistantes au feu (Rei 120) bien qu'elles soient montées dans un espace situé à l'extérieur. Chaque ascenseur peut transporter 13 personnes qui voyagent en toute détente à une vitesse de 1,6 mètres/seconde jusqu'à une hauteur de 30 mètres, et peuvent admirer la vue magnifique des environs (mer et collines, avec la république de Saint Marin au fond).
La réalisation du Centre Flaminio, construit par l'entreprise de Rimini Ge.Cos., a permis au chef-lieu de la Romagne de disposer d'une structure de qualité qui expérimente une série de solutions innovatrices en termes de construction et de technologie.

Massimiliano Colombo

Living Tomorrow

La Maison du Futur
Projet de Frank et Maurice Beliën
Abet Laminati

Le monde où nous vivons est en évolution constante. Notre société évolue à un rythme de plus en plus rapide et notre besoin d'avoir une meilleure qualité de la vie et de l'environnement croît sans cesse. L'humanité a pris conscience du problème de la protection de l'environnement, et la destruction des déchets, les économies d'énergie et l'utilisation de matériaux non polluants sont devenues des questions de première importance.
Nous voudrions tous que la Maison du Futur nous permette de vivre en suivant ces principes qui devraient caractériser une ère nouvelle.
Dans la Maison du Futur, nous voudrions pouvoir faire nos achats à domicile, recourir aux services bancaires par téléphone, communiquer de façon télématique avec notre bureau. Nous voudrions également que le cadre domestique soit plus sûr, plus sain et plus flexible. Si nous voulons vivre à l'avenir dans une telle maison, il nous faut nous en occuper dès maintenant.
En 1989, avec son père Maurice Beliën, architecte lui aussi, Frank Beliën se mit au travail pour élaborer le projet d'un milieu domestique de l'avenir, en cherchant à répondre à d'importantes questions telles que "Comment travaillerons-nous dans le futur? Comment voyagerons-nous? Quelle sera notre façon de vivre, de manger, de dormir... dans le futur?".
Des entreprises de premier niveau et opérant dans les domaines les plus divers furent alors contactées pour adhérer à ce projet auquel on leur demandait de ne fournir que des matériaux ou des services d'avant-garde. Aujourd'hui, le projet "Living Tomorrow" rassemble plus de 100 entreprises.
La première pierre a été posée le 5 octobre 1993 et l'inauguration officielle du complexe a eu lieu le 16 mars 1995. Living Tomorrow se trouve à Vilvorde, près de Bruxelles, où il couvre une surface totale de 15.000 mètres carrés dont 1120 sont occupés par la maison expérimentale. Il comprend quatre différentes zones : le Bureau du Futur, la Maison du Futur, le Complexe de l'Information où se trouvent un auditorium et des équipements d'exposition et, enfin, les Pavillons qui sont mis à la disposition des entreprises participant au projet.
Même si Architect Bureau Beliën doit être considéré comme étant l'entreprise promotrice, il convient de citer, parmi les 100 entreprises co-promotrices, Flemish Community, Tractebel (un groupe belge intervenant dans différentes entreprises publiques) et Abet Laminati.
Les activités de recherche qu'Abet Laminati mène dans le domaine des produits laminés la placent immédiatement au nombre des entreprises qui épousent parfaitement les principes de l'initiative "Living Tomorrow".
Abet Laminati a fourni les produits hpl suivants :
1. panneaux pour revêtements extérieurs MEG Standard et MEG Architettura ;
2. sols hpl Parqcolor ;
3. laminés Print hpl type "Metalli" ;
4. "Mobile bar" dessiné par Ettore Sottsass.
1. Pour les panneaux MEG Standard et MEG Architettura, l'architecte Frank Beliën a conçu spécialement un décor qui a été imprimé sur du laminé de base MEG dans l'établissement de Bra d'Abet Laminati. Les panneaux de revêtement mural MEG sont collés sur une structure en bois en utilisant les techniques de collage les plus avancées, puis ils ont été découpés dans des formes et des dimensions différentes pour habiller de couleur l'édifice de Living Tomorrow. Les laminés qui ont été utilisés sont des MEG Standard gris 871 et 475 et des MEG Architettura marron 799 ; leurs dimensions sont 3050 x 1300 x 6 mm et ils couvrent une surface totale de 40 mètres carrés.
Pour Frank Beliën, MEG est "un produit de revêtement mural absolument génial, car il se prête à n'importe quel type d'impression et de combinaison de dessin. Le collage invisible donne des résultats parfaits, aussi bien au niveau technique que sur le plan de l'esthétisme".
En plus des panneaux MEG imprimés spécialement, on a utilisé d'autres panneaux pour revêtir 120 mètres carrés sur les cinq piliers de soutien ; ils ont été montés sur des cadres en bois et vissés avec des vis-taraud peintes de la même couleur que le panneau. Le type utilisé est le MEG Architettura marron 799, de 3050 x 1300 x 6 mm de dimensions. Selon Frank Beliën, "esthétiquement parlant, les piliers de soutien en pierre siliceuse et en ciment n'étaient pas satisfaisants. Grâce à leur facilité d'installation et à leur gamme de couleurs, les panneaux MEG nous ont permis de résoudre brillamment le problème".
2. Sols hpl Parqcolor : il s'agit d'un revêtement de sol de 6,5 mm d'épaisseur, composé d'un panneau en fibre de bois revêtu des deux côtés d'un laminé décoratif à haute pression Print hpl et très résistant à l'usure. Finissage 9226 "Beech" dans la chambre d'enfants (14 mètres carrés), 9926 "I Tronchi" dans le bureau et dans le cercle de la "réalité virtuelle", 9937 "I Tronchi" dans le bar des pavillons. Sous de nombreux aspects, Parqcolor est un produit innovateur et il peut être joint à différents types de sol (linoléum et moquette) avec de minces profils flexibles. Dans le bar, il est inséré sur un sol à plaques de pierre et joint avec des joints à la silicone.
Les modalités d'utilisation sont pratiquement infinies. S'unissant aux nombreux finissages avec différentes reproductions d'essences, les excellentes performances (résistance à l'usure, aux rayures, etc.) et son aspect naturel et écologique font en sorte qu'il est possible d'utiliser Parqcolor de mille façons et pour les applications les plus variées (domestiques et semi-industrielles).
3. Print hpl "Metalli" : les finissages 877 Satinata et 877 Millerighe ont été utilisés pour le revêtement des meubles. Selon Beliën, le caractère "high-tech" du laminé Print "Metalli" s'adapte fort bien aux équipements très sophistiqués du complexe tout entier, même dans la salle de bains. Par ailleurs, le contraste entre ce laminé, dur et froid, et la chaleur du sol en carreaux est parfaitement compatible avec les différents contrastes chaud-froid, domestique-industriel, tranquille-affairé qui sont présents presque partout dans le complexe. D'autre part, le laminé "Metalli" a énormément simplifié la production des meubles en similaluminium.
4. "Mobile Bar" : un laminé spécial Print de la série "Serigrafia" a été produit expressément pour cet élément de l'ameublement qui a été dessiné par Ettore Sottsass.

Johnny Maris

Lumière pour l'Art

Un programme d'illumination des chefs-d'oeuvre de l'art italien
ENEL

Lumière pour l'Art est le programme lancé par l'ENEL pour témoigner que sa présence dans la réalité italienne n'existe pas seulement au niveau du monde de l'entreprise, mais aussi à celui de la culture.
Le vaste bagage acquis par l'entreprise en matière de technologie et d'expériences, au cours d'années où elle s'est livré à une activité spécifique, est mis à la disposition du public grâce à l'élaboration et à la réalisation de systèmes d'éclairage sophistiqués qui permettent de faire sortir de l'ombre quelques chefs-d'oeuvre du plus grand musée d'art et d'architecture du monde, l'Italie.
Mais Lumière pour l'Art n'est pas seulement lumière. Il s'agit d'un projet culturel et de vulgarisation d'une grande ampleur qui double la préparation des systèmes, effectués sur la base de projets d'éclairage spécialement conçus, d'une série d'importantes initiatives liées les unes aux autres, telles que la réalisation d'études et de documentations photographiques des oeuvres prises en considération, la publication d'ouvrages d'art et la production d'instruments multimédias (disques optiques compacts et programmes de réalité virtuelle) qui s'appuient sur les moyens technologiques les plus avancés en matière d'informatique et toujours selon la rigueur scientifique la plus absolue.
Lumière pour l'Art est donc un paquet de services complet et articulé, d'une haute teneur technologique et communicative qui peut être utilisée par toutes les structures institutionnelles, économiques et productives qui entendent contribuer à la mise en valeur d'un monument ou d'une oeuvre d'art de leur propre réalité locale, et ce, à l'aide d'interventions qui permettent de mieux l'admirer et qui en renforcent les contenus.
Conduite de pair avec les surintendances compétentes sur le territoire, afin de détecter les différentes oeuvres d'art à illuminer, et en symbiose avec les orientations du ministre de la Culture et de l'Environnement, la campagne a pris son envol en avril 1990 avec l'illumination des mosaïques de la basilique Saint-Marc de Venise. Les spectaculaires résultats de cette première intervention ont pleinement mis en valeur les marbres, les colonnes et surtout le complexe des mosaïques cachées dans la pénombre des grandes arcades.
En septembre 1990, le programme Lumière pour l'Art s'est intéressé au Dôme de Chieri, l'un des plus beaux exemples du gothique piémontais.
Ce fut ensuite le tour du Dôme de Gemona – symbole du long et très zélé travail de reconstruction qui suivit le tremblement de terre de 1976 – qui est illuminé depuis janvier 1991. Cette même année a également vu se réaliser les interventions de Piazza del Plebiscito à Ancône, de la basilique San Vitale et du Mausolée de Galla Placidia de Ravenne, ainsi que celles des basiliques supérieure et inférieure Saint-François d'Assise. L'éclairage de la place et de l'intérieur du Dôme de Spolète, celui de la basilique Saint-Laurent de Florence ainsi que celui des chambres peintes à fresque au Vatican par Raphaël, dans les appartements pontificaux de Jules II et de Léon X, ont été inaugurés en 1992 et 1993.
Ces réalisations sont également complétées par des travaux d'illumination "spot" – c'est-à-dire de courte durée – que l'on réalise avec des Modules d'éclairage : il s'agit d'un système innovateur et d'une technologie de pointe qui est utilisé à l'occasion de manifestations artistiques ou culturelles. Quelques exemples : l'illumination du Maschio Angioino, de Castel dell'Ovo, de Sant'Elmo de Naples et du splendide complexe monumental du palais de Caserte, à l'occasion de la rencontre des sept pays les plus industrialisés. N'oublions pas

celle des Forums impériaux et des Marchés de Trajan, et, plus récemment, l'illumination du Dôme et de la Place du Dôme de Lecce et du sanctuaire de Lorette, à l'occasion de la visite du Saint-Père.
Au-delà du fait que la diversification de leurs systèmes d'éclairage permet de mieux admirer différents chefs-d'oeuvre, ces interventions cherchent toutes à respecter pleinement l'atmosphère qui enveloppe l'oeuvre, à ne pas en suffoquer la perspective, à ne pas faire ressortir ses volumes de façon trop scénographique, et, surtout à ne pas altérer l'équilibre des effets de la lumière naturelle comme l'auteur les avait sentis à l'origine. Il a été prêté une attention particulière aux économies d'énergie électrique, et ce, grâce à l'utilisation appropriée et rationnelle des moyens technologiques, dans l'efficacité la plus complète.
Nous avons déjà souligné qu'il existe de nombreuses initiatives liées au programme considéré plus strictement au niveau du projet et des techniques d'illumination : leur but est toujours de mettre en valeur, de divulguer et faire mieux connaître le patrimoine artistique italien.
Pour ce qui est du domaine de l'édition, il faut rappeler les nombreux ouvrages d'art qui, à travers une documentation photographique détaillée, décrivent les opérations techniques effectuées et la richesse artistique du lieu de l'intervention. Parmi ces derniers, il convient de citer les deux prestigieux volumes qui ont été publiés à l'occasion de l'illumination de la basilique Saint-Marc de Venise.
La réussite de l'intervention menée pour l'illumination des basiliques Saint-François d'Assise a été marquée par la publication d'un volume fort intéressant. Une importance fondamentale est également prise par les ouvrages liés à l'opération qui a été menée au Vatican, à la Pinacothèque et dans les chambres peintes à fresque par Raphaël. Les dernières réalisations effectuées en matière d'édition sont consacrées à la Galerie Borghese et aux Forums impériaux. N'oublions ni les vidéos, ni les multivisions représentant les interventions les plus significatives qui ont permis, par exemple, d'explorer les fresques de Saint-François d'Assise dans leurs multiples perspectives.
Lumière pour l'Art recourt également aux techniques de l'informatique les plus avancées. La plus significative de ces dernières est certainement la réalité virtuelle qui est à même de recréer, grâce à l'électronique, des milieux et des situations, en saisissant le spectateur par des effets d'un réalisme extraordinaire. L'un des meilleurs exemples est "La ville de Giotto" : il s'agit d'un programme qui permet au spectateur de "pénétrer" dans les fresques que le Maître a peintes dans la basilique Saint-François d'Assise et de flâner dans une ville inexistante, reconstruite sur la base des images peintes par l'artiste. Tout aussi suggestif, le voyage dans la basilique Saint-Pierre a été accompli en suivant la merveilleuse perspective de la colonnade jusqu'à l'intérieur de l'église ; il permet également de visiter le noyau antique de Constantin qui a été reconstruit sur la base de très sérieuses recherches scientifiques. Ces applications sont possibles grâce à un système informatique extrêmement sophistiqué qui est le résultat d'études longues et complexes menées par l'ENEL.
Citons également la série de disques optiques compacts réalisée par l'ENEL pour exécuter un travail de vulgarisation intelligente de l'art, et ce, grâce à des programmes de visites d'expositions et de musées et à la restauration simulée de façon interactive de statues et de célèbres peintures.
La culture humaniste et la technologie se sont donc unies pour réaliser des travaux d'une grande importance.
Au fil des années, Lumière pour l'Art est donc devenue un promoteur et un transmetteur de culture digne d'autorité.

Aletto und Canaletto

Die Schlaftechnologie
Entwürfe von Paolo Piva und von Antonio Citterio
B&B Italia

Ein Zusammenhang, in dem der Ausdruck "Ergonomie" wirklich wesentlich ist, ist der Schlaf. Wenn man bedenkt, daß der Mensch ein Drittel seines Lebens schlafend verbringt, wird es klar, wie wichtig die Rolle des Bettes und seiner Bestandteile (Sprungrahmen, Matratze und Kissen) wird.
B&B Italia hat, im Einklang mit seiner Forschungs- und Erneuerungstradition einige von Paolo Piva und Antonio Citterio entoworfenen Bette hergestellt, worin die Suche nach neuen ästhetischen Lösungen mit der besten "Schlaftechnologie" gepaart worden ist.

Aletto

Das Bild schlichter Kraft und technologischer Strenge mit einer eigenen Note unterscheidet Aletto, von Paolo Piva entworfen.
Zwei Flügel aus Metallguß geben dem weichen Stoffnetz einen Schwung, der die Schultern beim Flug in die Traumwelt zu schützen weiß. Mit größter Einfachheit vereint sich die dichte, und doch leichte und transparente Webart der Baumwolle mit dem strengen Glanz des Aluminiums. Die absolute Entscheidung für die einzige verfügbare Version erlaubt sich nur eine Variante - die Wahl zwischen Schwarz und Beige. Um jedoch den nächtlichen Gewohnheiten des Benutzers entgegen zu kommen, fehlt es nicht an der Möglichkeit, einen drehbaren Beistelltisch zu montieren, der seine Funktion auf diskrete, nützliche und lautlose Weise erfüllt. In einer gelungenen Verbindung schafft der kalte Nerv aus Aluminiumguß Spannung zum warmen, einhüllenden Kopfteil aus Baumwolle und besänftigt mit der weichen Rolle aus Stoff oder Leder, die nach Art einer Nierenstütze angebracht ist.

Canaletto

Die schlichte Linie zeichnet mit leichtem Schwung eine weiche, minimale Einfachheit nach. Das Geheimnis von Canaletto, von Antonio Citterio entworfen, liegt im Bogen des Kopfteils und des Rahmens, der, vielleicht zu Ehren des venezianischen Malers, die Erinnerung an traditionelle Gondelbetten in sich trägt.
Der Überzug der Rückestütze paßt sich perfekt zwischen Kopfteil und Matratze ein. Der optische Eindruck ist durch die abgestimmten Kombinationen der Stoffbezüge von einer weichen Harmonie. Und wenn man will, kann man die Eleganz dieser Betten vollkommen abziehen - die Bezüge sind beliebig untereinander austauschbar. Auch für die Bodenauflage gibt es verschiedenen Lösungen: Die Füße dieses Bettes können ebenfalls mit Leder oder mit Stoff bezogen sein, sie können einen bodenlangen Bezug haben oder auch auf Rollen stehen.

Castelli für Territorio Italiano

Dreidimensionalität im Dienste der Kunst
Castelli

Im vergangenen September bot Castelli ihren Kunden ein neues Ereignis, das ein konkretes Zeichen des Willens der Firma war, ihre Identität als auf Kultur und Innovation achtendes Unternehmen zu bestätigen.
Im hübschen Rahmen des Museo dei Navigli in Mailand wurde eine "Territorio Italiano" (italienische Landschaft) betitelte Ausstellung eröffnet, die zu einem Abschnitt führte, der völlig mit dem neuen System 3D von Castelli für Büros realisiert worden war. Dieses Produkt zeigt die Lösung für die traditionelle Antinomie zwischen Flexibilität und Einfachheit, indem es dank der Verwendung der ganz wenigen Elemente, aus denen es besteht, geeignet ist, einen Renaissancegrundriß wiederzugeben, wie er der italienischen architektonischen Kultur lieb ist.
Schränke/Wachtürme, die sich über Schreibtischen/Gärten schließen, Empfangstheken/Mauern, die die Geheimnisse der Arbeitsflächen/Paläste verbergen. Wenn man durch die Holztür tritt, die seit 1877 das Symbol der alten Kunsttischlerei Castelli ist, befindet man sich so in einer traummäßigen Atmosphäre, wo die alte Tradition mit den neuesten Technologien zu einer formalen und strukturellen Synthese verschmilzt, die Castelli zum Erfolg geführt hat.
Mit Hilfe der ausgefeiltesten elektronischen Technologie und einem unterhaltsamen Drehbuch erstellte die Firma einen Kurzfilm, in welchem man die Zusammenstellbarkeit des Systems 3D, die Linie und die Farben, die es in jedem Zusammenhang vollkommen machen, bewundern kann: vom großen Mehrzweckloft zum Repräsentationsbüro, vom Konferenzsaal zu traditionellen Räumen.
Das war der Rahmen, den Castelli für die Ausstellung "Territorio Italiano" vorbereitet hatte.
1993 forderte Giacinto di Pietrantonio zahlreiche zeitgenössische Künstler, die es gewohnt waren, dreidimensional zu arbeiten, auf, sich an einem anregenden Unternehmen zu versuchen, der Interpretation der italienischen Landschaft. Die Projektsphilosophie ist es, die idealen Bedingungen wiederherzustellen, welche die italienische Kunst in der Geschichte groß gemacht haben, wodurch somit dieses außerordentliche Land als Bühne/ Anregung für die Phantasie der Künstler angeboten wird.
Damit entsteht "Territorio Italiano", mehr als eine Ausstellung, eine universelle und manchmal transgressive Botschaft, die dem Publikum den Stand der zeitgenössischen Kunst vermitteln soll.
Castelli berstätigt mit dem Vorschlagen dieser neuen, um weitere bedeutende Werke bereicherten, Ausgabe der Ausstellung ihre Gabe für Originalität und Innovation bei vollem Respektieren der Tradition.

Dodicesima Notte

Das erste ganznatürliche Bettsystem
Entwurf von Luca Meda
Molteni & C.

Es hat sich ein neues Bewußtsein des gesunden Schlafens entwickelt. Die am besten geeignete Schlafzimmereinrichtung sieht heute Bettgestelle und Roste mit anatomischen Latten aus Holz vor, Matratzen und Kissen mit natürlicher Polsterung, Wäsche aus Naturstoffen. Das "umweltverträgliche Bett" muß so gebaut sein, daß es die Metallelemente auf ein Minimum reduziert, auch um negative Einflüsse von Magnetströmen auf den Körper zu vermeiden. Bei Molteni war man stets von der großen dem Schlafbereich beizumessenden Bedeutung überzeugt: dies beweist das Bett "Morna", das 1971 den allumfassenden Einsatz des Holzes auch für die Halterungen vorsah, die gewöhnlich aus Metall sind. Mit Dodicesima Notte, entwickelt von Luca Meda, ist die Tradition des praktischen Schlafambientes und die Anwendung eines guten Designs nicht mehr allein auf das Bett begrenzt, sondern erstreckt sich auch auf seine Umgebung, indem ein Programm geschaffen wurde, bei dem ausgestattete Kopfteile, verfahrbare Podeste, modulare Regaleinheiten und weiteres Zubehör eine persönlich gestaltete und bequeme Schlafzimmereinrichtung bilden. Jedes Teil kann im Laufe der Zeit der Grundstruktur hinzugefügt werden, so daß eine vielfältige Nutzung und eine fortschreitende Vervollständigung der Einrichtung ermöglicht wird: das Bett kann über ein Podest geschoben werden, ein neuer Nachttisch findet Platz am ausgestatteten Kopfteil oder es wird eine weitere Regaleinheit hinzugefügt. Die massiven Holzteile der Betten und des Zubehörs sind aus Buche und Nußbaum, das Schichtholz besteht aus am Ursprung kontrollierten Paneelen ohne Formaldehyd und wird mit atoxischem Kleber zusammengefügt; die Polsterungen sind aus Kokosfaser, Latex und Baumwolle, die Soffe aus reiner Baumwolle und Mischleinen. Für ein Ambiente, in dem die Nacht angenehm ist und der Schlaf gesund, und wo es bei Tag genauso entspannend und bequem sein kann, in Ruhestellung zu lesen, zu schreiben oder mit einem Freund zu telefonieren.

Das Bürozentrum Flaminio in Rimini

Externe Panoramaaufzüge als Trümpfe
Projekt von S. Piccioli und G. Perazzini
Sabiem

Das kürzlich eröffnete Büro- und Einkaufszentrum Flaminio stellt einen bedeutenden Requalifizierungseingriff im an der alten Via Flaminia gelegenen Stadtgebiet von Rimini dar.
Die Entscheidung für die Errichtung eines Qualitätsbaus als Archetyp einer künftigen Evolution des Bauprodukts brachte zunächst eine originelle architektonische Form hervor. Aus diesem Grund

bedienten sich die Projektingenieure ungewöhnlicher Volumen, Formen und Materialien. Das Gebäude wird durch die vertikale Trägerstruktur (vollkommen aus Stahl, direkt von den Fundamenten ausgehend), die durch spiegelnde Fassaden und Aluminiumbugelemente erzielte äußere Umhüllung und vor allem durch die vier außen liegenden Panoramaaufzüge von Sabiem gekennzeichnet.
Besondere Aufmerksamkeit für die baulichen Details wurde für die Panoramaaufzüge aufgewandt, deren Schacht sich völlig im Freien befindet, ohne den Schutz der üblichen transparenten Elemente. Infolgedessen wurden sämtliche Anschlußstellen verzinkt, um Korrosion zu vermeiden, und es wurden neue Türen für die Aufzugsschächte entworfen, um den speziellen Funktions- und Normenbedürfnissen zu entsprechen. Die Öffnungen sind nämlich hermetisch abgedeckt, um das Eindringen von Regenwasser zu vermeiden, besitzen aber auch Brandschutztüren (Rei 120), obwohl sie sich an einem im Freien liegenden Raum befinden.
Jeder Aufzug hat eine Tragfähigkeit von 13 Personen, die entspannt das prachtvolle Panorama der Umgebung bei einer Geschwindigkeit von 1,6 Metern pro Sekunde und bis zu einer Höhe von 30 Metern bewundern können (Meer und Hügel, mit der Republik San Marino im Hintergrund).
Die Realisierung des vom Rimineser Unternehmen Ge.Cos. erbauten Flaminio gab dieser bedeutenden Stadt der Romagna die Möglichkeit, über eine qualifizierende bauliche Struktur zu verfügen, die eine Reihe wirklich moderner konstruktiver und technologischer Lösungen erprobt.

Massimiliano Colombo

Living Tomorrow

Das Haus der Zukunft
Projekt von Frank und Maurice Beliën
Abet Laminati

Wir leben in einer in ständiger Änderung befindlichen Welt. Unsere Gesellschaft entwickelt sich in immer schnellerem Rhythmus, und unsere Ansprüche an eine bessere Lebens- und Umgebungsqualität steigen ständig. Die Menschheit wurde sich des Problems der Umweltverschmutzung bewußt, und Müllbeseitigung, Energieeinsparung und Verwendung von die Umwelt nicht belastendem Material wurden zu Fragen erstrangiger Bedeutung.
Wir alle möchten, daß man im Haus der Zukunft diesen Prinzipien gemäß, die eine neue Ära kennzeichnen sollten, leben kann.
Im Haus der Zukunft möchten wir die Möglichkeit haben, Einkäufe von zuhause aus zu tätigen, die Bankgeschäfte per Telephon zu erledigen, telematisch mit dem Büro verbunden zu sein. Wir möchten auch ein sichereres, gesünderes und flexibleres häusliches Ambiente.
Wenn wir in Zukunft in einem solchen Haus leben wollen, müssen wir bereits jetzt beginnen, uns damit zu befassen.
1989 begann der Architekt Frank Beliën gemeinsam mit seinem Vater, dem Architekten Maurice Beliën, am Projekt eines häuslichen Ambientes der Zukunft zu arbeiten, indem er versuchte, eine Antwort auf wichtige Fragen wie "Wie werden wir in Zukunft arbeiten? Wie werden wir reisen, leben, essen, schlafen... in Zukunft?" zu finden.
Auf den verschiedensten Gebieten tätige, führende Unternehmen wurden kontaktiert, um an diesem Projekt teilzunehmen, unter der Bedingung, daß sie innovatives Material oder Service lieferten. Heute zählt dieses Projekt, "Living Tomorrow", bereits über 100 Teilnehmerunternehmen.
Am 5. Oktober 1993 wurde der Grundstein gelegt, und am 16. März 1995 wurde der Komplex offiziell eröffnet.
"Living Tomorrow" befindet sich in Vilvorde in der Nähe von Brüssel auf einer Gesamtfläche von 15.000 Quadratmetern, von denen 1120 von dem Versuchshaus bedeckt sind, und besteht aus vier verschiedenen Abschnitten: das Büro der Zukunft, das Haus der Zukunft, der Informationskomplex mit Auditorium und Ausstellungsausrüstung, sowie Hallen zur Verfügung der teilnehmenden Unternehmen.
Auch wenn das Architect Bureau Beliën als das promotende Unternehmen anzusehen ist, dürfen von den über 100 teilnehmenden und co-promotenden Firmen Flemish Community, Tractebel (ein belgischer Konzern mit Interessen in verschiedenen öffentlichen Bereichen), sowie Abet Laminati nicht vergessen werden.
Die Forschungstätigkeit von Abet Laminati auf dem Gebiet der innovativen Produkte aus Schichtpreßstoff weist es sofort als perfekt den Prinzipien der Initiative "Living Tomorrow" entsprechendes Unternehmen aus.
Abet Laminati lieferte die folgenden hpl-Produkte:
1. Paneele für Außenverkleidung MEG Standard und MEG Architettura;
2. hpl-Böden Parqcolor;
3. Print hpl-Schichtpreßstoffe Typ "Metalli";
4. "Mobile Bar", entworfen von Ettore Sottsass.
1. Für die Paneele MEG Standard und MEG Architettura entwarf der Architekt Frank Beliën eigens eine Verzierung, die auf den Grundschichtpreßstoff MEG in der Anlage von Abet Laminati in Bra gestanzt wurde. Die Mauerverkleidungspaneele MEG werden unter Einsatz der modernsten Verleimtechniken auf eine Holzstruktur geleimt und wurden dann in verschiedene Formen und Abmessungen geschnitten, um das Gebäude von "Living Tomorrow" mit Farbe zu verkleiden. Die verwendeten Schichtpreßstoffe sind MEG Standard grau 871 und 475 und MEG Architettura braun 799 in den Abmessungen 3050 x 1300 x 6 mm für eine Gesamtfläche von 40 Quadratmetern.
Frank Beliën definierte MEG "ein absolut geniales Produkt für die Mauerverkleidung, weil es jede Art des Stanzens und der Farbkombination erlaubt. Die unsichtbare Verleimung ergibt sowohl auf technischer, als auch auf ästhetischer Ebene perfekte Resultate".
Neben den eigens gestanzten MEG-Paneelen wurden weitere Paneele verwendet, um 120 Quadratmeter auf fünf Stützpfeilern zu verkleiden, und wurden auf Holzrahmen montiert und mit selbsteinschneidenden, in der gleichen Farbe wie das Paneel bemalten, Schrauben befestigt. Die verwendete Art ist MEG Architettura braun 799 in den Abmessungen 3050 x 1300 x 6 mm.
Laut Frank Beliën "waren die Stützpfeiler aus Kieselstein und Zement vom Ästhetischen her unbefriedigend. Die MEG-Paneele ermöglichten uns mit ihrer einfachen Anbringung und Farbenvielfalt, das Problem brillant zu lösen".
2. hpl-Böden Parqcolor: Es handelt sich um einen den Boden abdeckenden Fußboden von 6,5 mm Dicke, der aus einem Holzfaserpaneel besteht, das auf beiden Seiten mit Hochdruck-Schmuck-Schichtpreßstoff Print hpl verkleidet und sehr verschleißfest ist. Ausarbeitung 9226 "Beech" im Kinderzimmer (14 Quadratmeter), 9926 "I Tronchi" im Studio und im Kreislauf der "virtuellen Wirklichkeit", 9937 "I Tronchi" in der Hallen-Cafeteria.
Parqcolor ist ein unter vielen Gesichtspunkten innovatives Produkt und kann mit Böden verschiedener Art (Linoleum und Spannteppich) mittels dünner, flexibler Einfassungen verbunden werden. In der Cafeteria ist es in einen Steinplattenboden eingefügt und mittels Silikonverbindungen angebracht.
Die praktisch unendlichen Verwendungsmöglichkeiten bewirken zusammen mit den vielen Ausarbeitungen mit verschiedenen Holzreproduktionen, den hohen Leistungen (verschleiß- und kratzfest, usw.) und dem natürlichen, ökologischen Aussehen, daß Parqcolor auf viele Arten und in den verschiedensten Bereichen (häuslich und halbindustriell) verwendet werden kann.
3. Print hpl Typ "Metalli": Die Ausarbeitungen 877 Satinata und 877 Millerighe wurden für die Verkleidung der Einrichtungen verwendet.
Das High-Tech-Aussehen des Schichtpreßstoffes Print "Metalli" eignet sich laut Beliën ausgezeichnet für die raffinierten Ausstattungen des gesamten Komplexes, sogar im Badezimmer. Außerdem ist der Kontrast dieses harten, kalten Schichtpreßstoffes zur Wärme der verfliesten Böden mit den verschiedenen Warm-Kalt-, Haus-Industrie-, Ruhe-Beschäftigung-Kontrasten, die im Komplex überall vertreten sind, perfekt kompatibel. Außerdem hat der Schichtpreßstoff "Metalli" die Produktion von Einrichtungen aus Aluminiumersatz äußerst vereinfacht.
4. "Mobile Bar": Für dieses von Ettore Sottsass entworfene Einrichtungsstück wurde eigens ein Spezialschichtpreßstoff Print aus der Serie "Serigrafia" erzeugt.

Johnny Maris

Licht für die Kunst

Ein Programm zur Beleuchtung der großen Meisterwerke italienischer Kunst
ENEL

Licht für die Kunst ("Luce per l'Arte") ist das von ENEL erarbeitete Programm, das ihre nicht nur unternehmerische, sondern auch kulturelle Präsenz innerhalb der italienische Realität bezeugen soll
Die umfangreiche technologische und erfahrungsmäßige Ausstattung, die von dem Unternehmen in Jahren spezifischer Arbeit erworben wurde, wird der Öffentlichkeit durch die Projektierung und Realisierung ausgefeilter Beleuchtungsanlagen zur Verfügung gestellt, um einige Meisterwerke des größten Kunst- und Architekturmuseums der Welt – Italien eben – dem Dunkel zu entreißen.
Doch ist Licht für die Kunst nicht nur Licht. Es handelt sich um ein großzügiges kulturelles Popularisierungsprojekt, das neben die Erarbeitung der Anlagen auf Grund eigener Beleuchtungsprojekte eine Reihe wichtiger, untereinander verbundener Initiativen stellt, wie die Durchführung von Untersuchungen und Photodokumentationen der untersuchten Werke, die Publikation von Kunstbänden und die Herstellung von multimedialen Instrumenten wie CD-ROM und Programme virtueller Realität, die sich der modernsten Informatiktechnologien bedienen – alles im Rahmen absoluter wissenschaftlicher Strenge.
Licht für die Kunst ist also ein komplettes, gegliedertes Servicepaket von hohem technischem und Kommunikationsgehalt, das von allen institutionellen, wirtschaftlichen und produktiven Strukturen verwendet werden kann, welche zur Aufwertung eines Bau- oder Kunstwerks ihrer örtlichen Realität durch Maßnahmen beitragen wollen, durch welche sie besser zu genießen sind, und die ihren künstlerischen Gehalt unterstreichen.
Die Initiative wurde – mit den gebietsmäßig zuständigen Superintendanzen abgestimmt, um die verschiedenen, zu beleuchtenden Werke herauszufinden, und in Übereinstimmung mit den Richtlinien des Ministeriums für Kultur- und Umweltgüter – im April 1990 mit der Beleuchtung der Mosaike der Markuskirche in Venedig gestartet. Die spektakulären Ergebnisse dieser ersten Arbeit führten den Marmor, die Säulen und vor allem den im Halbdunkel der großen Bögen halb verborgenen Mosaikkomplex in ihrer ganzen Schönheit vor.
Dann wurde das Programm von Licht für die Kunst im September '90 mit dem Dom von Chieri fortgesetzt, einem der bedeutendsten Beispiele der Gotik im Piemont.
Dann wurde im Januar '91 der Dom von Gemona, das Symbol der zehnjährigen, fleißigen Wiederaufbauarbeit nach dem Erdbeben von 1976 beleuchtet. Im selben Jahr folgten die Maßnahmen bezüglich Piazza del Plebiscito in Ancona, Basilika San

Vitale und Mausoleum der Galla Placidia in Ravenna und der Oberen und Unteren Basilika San Francesco in Assisi aufeinander. 1992 bzw. '93 wurden die Beleuchtungsanlagen für den Domplatz und das Innere des Doms von Spoleto, für die Pinakothek des Vatikans, die Basilika San Lorenzo in Florenz und die von Raffael in den Papstgemächern Julius II. und Leo X. im Vatikan mit Fresken bemalten Räume eingeweiht.
Umringt werden diese Einsätze von sogenannten Spotbeleuchtungen, das heißt, daß sie von kurzer Dauer sind und mit Beleuchtungsmodulen erfolgen, einem innovativen High-Tech-System, das gelegentlich künstlerischer oder kultureller Ereignisse verwendet wird. Beispiele dafür sind die Beleuchtung des Maschio Angioino, von Castel dell'Ovo und Sant'Elmo (alle in Neapel) und des prachtvollen Monumentalbaus des Königsschlosses in Caserta gelegentlich des G7-Summits, der Kaiserforen und Trajanmärkte in Rom und vor kurzem des Doms und Domplatzes in Lecce und des Heiligtums von Loreto gelegentlich des Besuchs des Heiligen Vaters.
Leitmotiv all dieser Arbeiten war neben dem natürlichen Bedürfnis nach Verbesserung des Genusses der einzelnen Meisterwerke durch die Diversifizierung der Beleuchtungsanlagen immer auch die volle Beibehaltung der das Werk umgebenden Atmosphäre, das heißt, die Perspektive nicht zu verflachen, die Volumen nicht zu stark szenographisch herauszuheben und – vor allem – das Gleichgewicht der Effekte des natürlichen Lichts, wie es ursprünglich vom Künstler gefühlt wurde, nicht zu verändern. Besondere Aufmerksamkeit wurde einem mäßigen Energieverbrauch gewidmet, welches Ziel durch eine passende, rationelle Gestion der Technologie im Rahmen maximaler Effizienz erreicht wurde.
Zahlreich sind, wie gesagt, die mit dem eigentlichen Projekts- und beleuchtungstechnischen Programm verbundenen Initiativen, welche alle das künstlerische Vermögen Italiens aufwerten, verbreiten und gründlicher bekannt machen sollen.
Auf verlegerischem Gebiet ist an die zahlreichen Kunstbände zu erinnern, die mittels detaillierter Photodokumentation die erfolgten technischen Arbeiten und den künstlerischen Reichtum des betroffenen Werks belegen. Zu diesen gehören zwei prestigereiche Bände, die gelegentlich der Beleuchtung der Markuskirche in Venedig erschienen.
Auch der gelungene beleuchtungstechnische Eingriff in den Basiliken San Francesco in Assisi wurde durch die Publikation eines bedeutungsvollen Bandes besiegelt. Grundlegende Bedeutung kommt auch den Bänden in Zusammenhang mit der in der Pinakothek des Vatikans und den von Raffael mit Fresken ausgemalten vatikanischen Stanzen erfolgten Arbeit zu. Die jüngsten verlegerischen Taten sind der Galleria Borghese und den Kaiserforen in Rom gewidmet. Nicht zu vergessen sind auch die mit den bedeutendsten Maßnahmen verbunden Videos und Multivisionen, wie bei den Fresken von San Francesco in Assisi, die in ihren vielfachen Perspektiven erforscht wurden.
Licht für die Kunst bedient sich auch der modernsten Informatiktechnologien. Die bedeutendste darunter ist zweifellos die virtuelle Realität, die imstande ist, elektronisch Räume und Situationen zu schaffen und den Betrachter mit Effekten eines erstaunlichen Realismus' miteinzubeziehen. Ein Beispiel dafür ist "Die Stadt Giottos", ein Programm, mit dessen Hilfe der Besucher die in der Basilika San Francesco in Assisi vorhandenen Fresken des Meisters "betreten" und in einer inexistenten Stadt herumschlendern kann, die auf Grund der von dem Künstler gemalten Bilder rekonstruiert wurde. Ebenso zauberhaft ist die Reise in den Petersdom, die dem Weg der wundervollen Perspektive des Säulengangs bis ins Kircheninnere folgt, mit der Möglichkeit der Besichtigung des auf Grund sorgfältiger wissenschaftlicher Studien rekonstruierten alten Kerns aus den Zeiten Kaiser Konstantins. Diese Anwendungen bedienen sich eines äußerst raffinierten Computersystems, dem Ergebnis langer, komplexer Untersuchungen seitens ENEL.
Ebenso wichtig ist die Erarbeitung einer Reihe von CD-ROM, die von ENEL produziert wurde und auf die intelligente Verbreitung der Kunst mittels eigener Programme für den Ausstellungs- und Museumsbesuch und auf interaktive Weise simuliertes Restaurieren von berühmten Statuen und Gemälden ausgerichtet ist.
Humanistische Kultur und Technologie verschmelzen somit in Realisierungen von großer Bedeutung. Licht für die Kunst wurde mit den Jahren zu einem angesehenen Förderer und Verbreiter von Kultur.